Social Thermodynamics

Zhi Cheng

Social Thermodynamics

An Interdisciplinary View

PETER LANG
Lausanne • Berlin • Bruxelles • Chennai • New York • Oxford

Library of Congress Cataloging-in-Publication Data

Names: Cheng, Zhi, author.
Title: Social thermodynamics : an interdisciplinary view / Zhi Cheng.
Description: New York : Peter Lang, [2023] | Includes bibliographical references.
Identifiers: LCCN 2023007366 (print) | LCCN 2023007367 (ebook) |
ISBN 9781636670683 (hardback) | ISBN 9781636670669 (ebook) |
ISBN 9781636670676 (epub)
Subjects: LCSH: Social systems. | Social conflict. | Thermodynamics.
Classification: LCC HM701.C43 2023 (print) | LCC HM701 (ebook) | DDC
301–dc23/eng/20230331
LC record available at https://lccn.loc.gov/2023007366
LC ebook record available at https://lccn.loc.gov/2023007367
DOI 10.3726/b20738

Bibliographic information published by the **Deutsche Nationalbibliothek.**
The German National Library lists this publication in the German
National Bibliography; detailed bibliographic data is available
on the Internet at http://dnb.d-nb.de.

Cover design by Peter Lang Group AG

ISBN 9781636670683 (hardback)
ISBN 9781636670669 (ebook)
ISBN 9781636670676 (epub)
DOI 10.3726/b20738

Published by Peter Lang Publishing Inc., New York, USA
info@peterlang.com - www.peterlang.com

This publication has been peer reviewed.

Contents

List of Figures

List of Tables

Preface

Since the 19th century, sociological theory has developed rapidly. In particular, the development of science led to the application of Comte's positivism to sociology (Comte, 1858), which promoted sociology from an empirical theory to the level of social science. The development of sociological theory in turn promotes the development and change of the whole society. The creation of social conflict theories of Marx (Collins, 1975) led to the vigorous rise of the international communist movement. Dewey's social psychology theory (Deway, 1929) has caused changes in the field of education, promoting education from teacher-centered teaching to student-centered teaching (Yun, Li, Fu, & Liu, 2018). And Edgar Morin's complex thought theories (Morin, 1992) have promoted the development of social democracy and pluralism. From the impact of these sociological theories on social change, we can see that sociological theories occupy a very important position in the social sciences and are the foundation of the entire social sciences.

From the perspective of the theoretical development process of sociology, the theory of sociology has gradually developed from early philosophical thinking to the application of positivism theory, and then to the emergence of conflict theory, structural functionalism, symbolic interaction theory, and complex thinking, sociological theory has experienced a development process from macro to micro. The various important achievements in sociology so far seem to be mainly concentrated in the micro-field, which is mainly influenced by the gradual expansion of the research objects of natural science, especially natural science from the macrocosmic level to micro quantum mechanics, molecular biology, etc. For example, since the 1960s, Berger, Luckmann and Schutz's Social Construction Theory (Berger & Luckmann, 2016; Muzzetto, 2015), Mead, Blumer's social behaviorism and symbolic interactionism (Blumer, 1980), Goffman's Dramaturgical approach (Sarbin, 2003) and other micro-sociological theories have a great influence. Sociology has also begun to pay attention to the study of complex structures of societies at the microscopic level (Montuori, 2004; Morin, 1992).

However, in the 20th century, the field of macrosociology has not developed so fast. Especially since the collapse of the Soviet Union in the early 1990s, the validity of some macrosociological theory has been met with a degree of skepticism. This is also one of the reasons why the study of macro sociological theory has stagnated.

On the other hand, the lack of a newer macro-level sociological theory will bring more problems to modern social practice. Since the industrial revolution in the 18th century, many new problems have appeared in human society. Including the impact of human activities on climate change (National Academy of Sciences, 2020), the outbreak of two world wars, the invention and manufacture of nuclear weapons, the imbalance of world development, the impact of the rapid development of information technology on traditional society and so on. Among various problems, the biggest impact on human society is the global climate change caused by the human activities. Before industrialization, humans were only one component of the entire social system. The energy absorbed and consumed by human activities is basically the same as that of animals and plants in the earth ecosystem. After hundreds of thousands of years of development, human activities are also very effectively integrated into the entire earth's ecosystem. But since the industrialization of society, human activities began to have a huge impact on climate change, and even began to show the ability to dominate global climate change.

The ecological balance of the earth has been disrupted. Therefore, the social system will inevitably include animals and plants other than humans and the entire material cycle of the earth. Due to the lack of a globe change perspective, the existing sociological theories only look at the contradictions and conflicts of these issues from the perspective of human beings or society itself, and as a result, it has been unable to grasp the root causes of these social problems. Naturally, there is a lack of effective solutions, resulting in many social problems that are difficult to solve in modern society. Such as the uncontrollable rise in atmospheric temperature, the seemingly inevitable conflict between Russia and Ukraine, the effective control and destruction of nuclear weapons, etc.

Therefore, in the absence of an effective theoretical guide for studying social issues at a very macro level, such as the level of global change, each country may face these problems in a way that is beneficial to it. The end result is more problems, making it difficult to effectively implement long-term measures that can curb the rapid rise in global temperatures to some extent.

For example, during Russia's war in Ukraine in 2022, in response to the lack of natural gas, some European countries have begun to restart coal-fired power generation (Kappeler & John, 2022). In other strife, the climate agreement simply cannot be effectively implemented. Some countries have withdrawn from relevant climate agreements at will (Kann, 2019), while others suspend climate cooperation between countries (Yeung, 2022). Even in the summer of 2022, the average temperature in many cities of the world exceeds the historical record, which does not seem to attract enough attention to the seriousness of the problem of greenhouse gas emissions.

Of course, some theories related to sociology may also have various problems themselves. For example, Malthus's population theory (Weir, 1991) holds that population growth is exponential, and to suppress this population growth requires war and famine. But in fact, from the perspective of human development history, although the short-term population and the reduction of human activities brought about by the war do exist, after the war, the population will have a stronger revenge growth process. For example, the emergence of the baby boom after World War II was the period of the fastest population growth in human history. After decades of long-term peaceful development, the population growth rate of many developed countries has slowed down, and even negative population growth has occurred. This also reflects that the previous sociology-related theories may encounter more difficulties when they are applied in practice today.

In addition to the lack of macroscopic research perspective, the previous sociological theory also has a deficiency, which is reflected in the insufficient mathematical tools used. The vast majority of sociological theory is based on qualitative analysis. The advantage of using qualitative analysis is that it has a high generality and can better generalize some very complex issues. Qualitative approaches, however, introduce conceptual uncertainty. Since everyone has their own cultural background and their own way of understanding, there may be very large differences in the understanding of the same concept.

In turn, we can refer to quantitative analysis methods in the field of economics. Since the importance of quantitative analysis has been recognized for a long time in the field of economics, very effective mathematical tools have been established long ago to solve economic problems. This has also contributed to the development of the modern economy. The continuous development of the global economy has also promoted world peace and stability.

Therefore, if it is possible to use mathematical tools to establish quantitative models similar to economics in sociological theory, I believe that it can be helpful to eliminate the uncertainty of various concepts and to analyze various problems existing in human society.

For this reason, this book will look at social issues from a macro perspective of global change. This also means that human society not only has the interaction of internal social individuals, but the entire social system is also closely related to the global environment in which it exists. These global environments include global geological and atmospheric systems and ecosystems composed of biodiversity. Human society not only has a self-organizing process maintained by the internal dynamic system of the society, but also continuously absorbs energy from the sun, and also interacts with various subsystems in the global environment.

Historical document data show that when the global temperature is at a high enough level, the entire human civilization is in a very prosperous era. When the temperature of the entire earth enters the cold ice age, various conflicts and wars will occur frequently in human society. Of course, in modern society, with the development of modern technology, a large amount of fossil energy stored under the earth's crust has been excavated and burned, which will generate additional heat in the entire earth's environment. At the same time, the burning of fossil fuels releases a large amount of greenhouse gases such as carbon dioxide, which further causes the earth's atmospheric temperature to rise at an unprecedented rate. This rapid rise in atmospheric temperature has brought great uncertainty to the development of human society. In this monograph on the thermodynamics of society, some of these issues will be discussed at a relatively large length.

Therefore, in terms of content arrangement, Chapter 1 of this book provides an overview of some basic knowledge of physics and thermodynamics. In particular, the various parameters that need to be used in subsequent chapters are listed in the first section of this chapter, so that they can be easily referenced in subsequent chapters.

The development of thermodynamics and the development of sociology are contrasted in Section 1.2. This contrast allows us to see the relationship between the development of thermodynamics and the development of sociology. This also allows us to deeply understand why it is necessary to introduce the methodology of thermodynamics in sociology to solve social problems.

Because the structure of the social thermodynamic system is very complex. Most of the various factors involved are qualitative concepts, and some concepts even involve the meaning of literature and art. In order to meet the quantitative requirements in subsequent chapters, some microscopic concepts need to be scientifically defined. Therefore, the analysis in Chapters 2 and 3 has gradually risen from the research on the underlying information transmission and the composition of neural network cells to the research on how neural networks process information and human cognitive processes. These are relatively microscopic system structures, which directly determine the macroscopic social structure. Because in the social thermodynamic system, these microscopic system structures lead to great differences in the interactions between different social individuals. And those different interactions form the basic social members of different internal structures. Ultimately, the interaction between the basic social members of these different structures forms different types of societies.

Chapter 2 proposes a division of three different types of societies, which are the focus of this book. Therefore, the content of Chapters 2 and 3 of this book is indeed very necessary. However, if you want to quickly understand the calculation of social thermodynamic systems and the laws of phase transitions, you can skip Chapters 2 and 3 and read Chapter 4 directly. Because Chapter 4 begins with an analysis of the general properties of social thermodynamic systems, the reliance on the content of the previous two chapters is less strong. This is also consistent with the non-linear way some academic writings use to arrange content. If you run into problems reading Chapters 4, 5, and 6, you can always go back to Chapters 2 and 3 for comparison.

Of course, if you have enough patience and time, it is recommended to read directly in the order of the chapters. In this way, the main knowledge of social thermodynamics can be comprehensively and systematically understood.

Chapters 4, 5, and 6 apply the general laws of thermodynamics to social thermodynamic systems. This involves the quantification of various parameters. In Chapter 4, energy, social pressure, social space, and social temperature in thermodynamics are quantified. These quantifications are combined with actual physical processes. Therefore, the quantified sociological parameters have the same dimension as the physical thermodynamic system parameters. For example, the social temperature after quantification is comparable to the physical temperature of the environment. This also means that if the social temperature of the social thermodynamic system is lower than the physical temperature of the environment, according to the zeroth law of thermodynamics, the social thermodynamic system is more likely to absorb energy from the environment. This will promote the social thermodynamic system to a more stable state. And if the social temperature of the social thermodynamic system exceeds the ambient temperature, it means that it will be more difficult for the social thermodynamic system to absorb energy from the environment. The whole society will be in an unstable state.

In Chapter 5, these quantified social thermodynamic parameters are related to each other through the equation of state. By comparing the physical thermodynamic system, we can know that there are isobaric, isovolumetric and adiabatic processes in the social thermodynamic system. The energy absorbed and released by different social thermodynamic processes and the changes in social temperature are not the same. Solving the equation of state can be used to explain some important social phenomena. For example, the number of animal groups is limited, the long holiday effect, etc.

Chapter 6 discusses a very important social thermodynamic phenomenon, which is the phase transition. The existence of phase transition often means drastic changes in society. Of course some phase transitions are good, others can be catastrophic. By solving the critical parameters of a thermodynamic system, we can learn under what conditions a society is likely to undergo a phase transition and avoid catastrophe. In addition, in the above chapters, in order to avoid complicated mathematical formulas interfering with readability, some contents that have more mathematical formulas or have little influence on the understanding of concepts are listed in the form of "Math further". If it does not affect the understanding of the content, it can be skipped when reading.

Chapters 7, 8, and 9 respectively analyze the family-oriented society, the social-oriented society and the balanced society. Each of these three chapters has a very detailed analysis of the origins of each society. This could explain some of the things we've been confused about in the past. For example, why are there so many different types of societies and cultures in this world? Why are there conflicts between different types of social cultures? How to resolve these conflicts? and many more. These should all be able to obtain corresponding answers and solutions from the theory of social thermodynamics.

Finally, thanks to the scholars who have peer-reviewed this book. Their very detailed analysis and recommendations have been very helpful in improving the quality of this book.

Zhi Cheng

Nov. 2022. Guangzhou, China.

References

Berger, P., & Luckmann, T. (2016). The social construction of reality. In *Social theory re-wired* (pp. 110–122). Routledge, New York.

Blumer, H. (1980). Mead and Blumer: The convergent methodological perspectives of social behaviorism and symbolic interactionism. *American Sociological Review*, 45(3), 409–419.

Collins, R. (1975). *Conflict sociology*. New York: Academic Press.

Comte, A. (1858). *The positive philosophy of Auguste Comte*. New York: Blanchard.

Deway, J. (1929). *Characters and events: Popular essays in social and political philosophy* – Volume I. New York: Henry Holt and Company.

Kann, D. (2019, November). US begins formal withdrawal from Paris climate accord. *CNN*. https://edition.cnn.com/2019/11/04/politics/trump-formal-withdrawal-paris-climate-agreement/index.html.

Kappeler, I., & John, T. (2022, June). Germany to fire up coal stations as Russia squeezes gas supply. *CNN*. https://edition.cnn.com/2022/06/19/energy/germany-russia-gas-supplies-winter-intl/index.html.

Montuori, A. (2004). Edgar Morin: A partial introduction. *World Futures*, 60(5–6), 349–355.

Morin, E. (1992). From the concept of system to the paradigm of complexity. *Journal of Social and Evolutionary Systems*, 15(4), 371–385.

Muzzetto, L. (2015). Schutz, Berger and Luckmann. The question of the natural attitude. *SocietàMutamentoPolitica*, 6(12), 245–277.

National Academy of Sciences. (2020). *Climate change: Evidence and causes: Update 2020*. Washington, DC: The National Academies Press. https://doi.org/10.17226/25733.

Sarbin, T. R. (2003). *The dramaturgical approach to social psychology: The influence of Erving Goffman*. American Psychological Association.

Weir, D. R. (1991). Malthus's theory of population. In *The world of economics* (pp. 401–406). Palgrave Macmillan, London.

Yeung, J. (2022, August). China suspends cooperation with US on range of issues, sanctions Pelosi over Taiwan trip. *CNN*. https://edition.cnn.com/2022/08/05/asia/nancy-pelosi-taiwan-china-tokyo-intl-hnk/index.html.

Yun, C., Li, S., Fu, Y., Liu, C., & Liu, F. (2018, January). Study on roles of teachers in applied Medical Colleges under the Student-centered Idea. In 2017 5th International Education, Economics, Social Science, Arts, Sports and Management Engineering Conference (IEESASM 2017) (pp. 138–141). Atlantis Press.

Ethical considerations for this book

Since this book involves the relationship between human society and the natural environment, as well as the survival and development of human society itself, when writing this book, I also noticed the possible impact of the content of this book on existing social forms and cultures. Therefore, I believe this will involve some ethical issues. In order to prevent the publication of this book from adversely affecting various ethical requirements and different cultural values, the following statement is hereby made:

1. This book is a monograph on social issues from the perspective of global change, which belongs to a kind of sociological theory. Like other sociological theories, this book inherits the development of sociological theory since the 19th century. But compared with previous sociological theories, this perspective is more macroscopic. It is because, considering global development issues, human activities have begun to affect the changes in the global environment. For example, human activities can even lead to an abnormal increase in global atmospheric temperature,

which brings great uncertainty to human society. How to make humans understand these new problems and uncertainties is the basic starting point of this book. Of course, in the process of writing this book, it also draws on the knowledge of many other disciplines. Therefore, this book can use new tools, such as thermodynamic calculation tools, to explore many deep-seated problems in society.

2. The content of this book does not initiate, nor does it lead to, some possible social movement or revolution. On the contrary, through the discussion of all aspects of human society, we can look at the social problems from a higher level. The use of some mathematical methods is more helpful for us to see the relationship between various factors that affect social development. Even if there are factors that do not meet the needs of social development, the discussion in this book will help us to promote a smooth transition of society in a more gradual way.
3. The book touches on ethnic and social issues when discussing different social formations. But there is no bias in dealing with the problems. The author respects all cultures and social formations. This book does not suggest which social form is superior.
4. The content of this book aims to reveal the laws of human society, and the purpose is to solve the problems existing in human societies. This includes democracy, equality, the rule of law, etc. It also involves issues such as social development trends and the impact of global climate change on human society.
5. Because this book has used the knowledge of thermodynamics, it has revealed the laws contained in various social forms, which involve both human society and animal society. It is believed that by adjusting those thermodynamic parameters that are not suitable for society, the severity of human-social conflict can be alleviated, and the conservation problem of animal society can also be helped.
6. Although this book studies macroscopic social issues, it also pays attention to the relationship between individuals and social development. Such as the interaction between individuals, the cognitive development process of individuals and so on.

Therefore the content of this book is also believed to help individuals understand their place and role in society.

7. This book mainly uses the knowledge of thermodynamics to solve social problems. However, when dealing with it, it does not completely equate human society with the physical world. This is mainly manifested in: (1) This book does not use fully accurate quantitative calculations in physical thermodynamics. Many parameters are approximate estimates. (2) This book combines quantitative and qualitative analysis. Although it appears to be a quantitative calculation in form, it mainly focuses on qualitative analysis. (3) Fully aware of the differences between human or animal society and matter systems. (4) Pay attention to the differences between different individuals and systems in a society. (5) The conclusions in the book are for reference only, and provides a possible basis. However, specific decision-making still requires human beings to conduct specific analysis of specific issues based on individual and social differences.

1

Thermodynamics and sociological fundamentals

1.1 Thermodynamic basics

1.1.1 Interactions between matter

Thermodynamic is an important branch of physics. Like other branches of theory, thermodynamic is actually based on the mechanics of physics. These mechanical foundations mainly include four major mechanics, namely Newton's classical mechanics, quantum mechanics, electrodynamics and relativity.

If you are dealing with general room temperature and macro-level problems, such as the properties of gases, liquids, and solids, all problems can be solved by using classical mechanics. Because under this condition, the motion law of gas molecules follows the Maxwell-Boltzmann distribution (Pathria & Beale, 2011).

But if it is at an extremely low temperature or at a more microscopic level, the mechanical basis on which thermodynamic relies at this time is quantum mechanics. At this time, the laws followed by molecular motion are mainly Bose-Einstein or Fermi statistical distributions.

If it enters the macroscopic level and reaches the level of the universe, the laws of thermodynamic at this time must be based on relativistic mechanics. For example, studying the thermodynamic properties of black holes is related to general relativity at this time.

The mechanical knowledge that is the basis of thermodynamic is mainly used to describe the interaction between molecules or matter. There are four main types of interactions known so far, namely gravitational interaction, electromagnetic interaction, weak interaction and strong interaction.

However, modern physical theories have demonstrated that the electromagnetic and weak and strong interactions are actually unified (Meyers, 2003). It can be described by a theory of interaction. In this way, when we deal with the molecular interactions at the bottom of thermodynamic, we only need to consider electromagnetic interactions and gravitational interactions. If the general macroscopic level or the thermodynamic system on the scale of human society is involved, the interaction at the molecular level only needs to consider the electromagnetic interaction and can solve almost all problems. At the macroscopic level, when dealing with macroscopic thermodynamic parameters, gravitational interactions can also be involved at this time. For example, the earth's atmosphere can form atmospheric pressure under the action of gravity.

For gravitational interactions, currently there are two main theories dealing with gravitational interaction. One is Newton's theory of gravity. The other is Einstein's General Theory of Relativity. Of course the two theories are essentially the same. With proper approximations, general relativity can be turned into Newton's theory of gravity.

In Newton's theory of gravity, it is emphasized that there will be gravitational interaction between objects with mass. The magnitude of this gravitational force can be expressed using the following formula

$$F = G\frac{Mm}{r^2} \tag{1-1}$$

where G is the gravitational constant, M and m are the two different masses, and r is the distance between the two masses. F represents the gravitational force created between these two masses.

In Newton's theory of gravity, time and space are flat. The generation of gravity is an important property of matter.

General relativity, on the other hand, argues that gravity arises as a result of the curvature of space-time. Due to the curvature of space-time, matter with mass or energy will move accordingly due to uneven space-time.

General relativity uses a space-time metric to express how curved space-time is. The space-time metric is a second-order tensor that can be represented by a matrix. Usually the notation $g_{\alpha\beta}$ is used to denote a metric. The subscript $\alpha\beta$ corresponds to the two components of the metric tensor matrix. Since space-time is four-dimensional, $\alpha,\beta = 0,1,2,3$. The 0 dimension represents the time dimension, and 1, 2, and 3 correspond to the three-dimensional spatial components of x, y, and z, respectively.

If space-time is flat, it is usually represented by $\eta_{\alpha\beta}$. This flat space-time metric is called the Minkowski metric. Considering that the theory of general relativity is indeed quite complicated, it can only accurately deal with spherically symmetric gravitational fields. Therefore, a four-dimensional coordinate system called "sphero-cylindrical coordinates" is most commonly used in general relativity. The spherical coordinates correspond to a three-dimensional spherically symmetric space.

For the space-time with spherically symmetric gravitational field, the closer to the gravitational field, the slower the time and the shorter the radial length.

Unlike Newtonian mechanics, the gravitational theory of general relativity can calculate the curvature of space-time. Especially under very strong gravitational conditions, the curvature of space-time becomes a completely closed state, and the so-called black hole appears at this time. If a black hole appeared, then space-time would be divided into two parts. One is the space-time outside the black hole. The outer space-time is a normal space-time, of course, the space-time itself is bent by the action of gravity. But all matter can still be observed. If there is enough energy, these matters can still fly out of very strong gravitational regions. But if it enters the interior of the black hole, the curvature of space-time at this time will make it impossible for all matter and even light to fly out of the black hole boundary. The boundary of this black hole is called the "event horizon" of the black hole (Cheng, 2021).

Unlike the gravitational interaction, the electromagnetic interaction mainly involves the interaction between two different electric charges. The interaction between charges follows Coulomb's law, and the form of the calculation formula is basically consistent with Newton's law of gravity.

$$F = k\frac{q_1 q_2}{r^2} \tag{1-2}$$

where k is the Coulomb constant and q_1 and q_2 are two different charges. Compared with Newton's law of gravitation, we can see that the interaction of electric charges also follows the law of inverse square. This actually reflects some kind of intrinsic link between gravity and electromagnetism.

In classical physics, both gravitational and electromagnetic interactions seem to be able to interact instantaneously and over long distances. That is to say, as long as two masses or two electric charges exist, there can be instant interaction between these two objects or electric charges, just like the air grabbing in magic. Such an interaction mechanism is physically difficult to calculate, especially in quantum field theory. It is also against the laws of physics. As Einstein pointed out (Einstein, Podolsky, & Rosen, 1935), in physics, if there are interactions of "Spooky Action at a Distance", then the process may not be a physical one.

To solve this problem, the concept of virtual or intermediate particles was introduced in quantum field theory. This means that if there is to be an interaction between two different charges, then some kind of intermediate particle must be transferred between the two charges.

The virtual or intermediate particles delivered by different interactions are different. The intermediate particles delivered by electromagnetic interactions are virtual photons. The intermediate particle transmitted by the strong interaction is the gluon. Weak interactions transmit Z and W bosons. Of course, in quantum field theory, it has been proved that these three interactions are actually the same interaction, and they can all be unified into electromagnetic interaction.

As for the gravitational interaction, there are also some theories that consider the existence of graviton, so these theories speculate that the gravitational interaction is carried out by exchanging the graviton. However, both in theory and in practice, this graviton's theory has great difficulties. In theory, if there is a graviton, the gravitational interaction is too weak, about 10^{-40} times that of the electromagnetic interaction, so if there is a graviton, this graviton will theoretically have a very, very small energy. Whether such a small energy can be called quantum still has many theoretical obstacles in physics. Because unlike mathematics, there should be no such measure as infinitesimal in physics. In addition, in practice, it is precisely because the theoretically predicted graviton energy is too small, so far it is impossible for any method to detect graviton. Of course, it has been proved that gravitational waves exist, so gravitational interaction can still be achieved by exchanging gravitational waves.

1.1.2 Feynman diagram representation of interactions

In order to more clearly express the interaction between matter or charges, Feynman used a very simple way to express this interaction. This is called a Feynman diagram (Veltman, 1994). The form of this diagram is very intuitive, and by setting some additional rules, Feynman diagrams can also be used to do some calculations. This makes Feynman diagrams very widely used in quantum field theory.

The rules of Feynman diagrams are actually quite simple. Although the space-time we are in now is four-dimensional, including one-dimensional time and three-dimensional space, but in the process of interaction, particles can actually be represented by one-dimensional position movement. Therefore, we can use a two-dimensional space-time to represent the interaction between two objects. In this two-dimensional space-time, one of the dimensions is time, which reflects the time dependence before and after particle interaction. The other dimension is space, which reflects the change in position of the two particles after they interact.

Since we now know that in the physical world, anti-matter is relatively common. Physically, the relationship between anti-matter and matter is opposite. In order to show such an important distinction, a very clever approach is adopted in Feynman diagrams, which is to use the direction of the arrow to indicate the positive and negative characteristics of the matter.

Therefore, in Feynman diagrams, a matter or particle actually corresponds to a line (rather than a point). This also reflects that in the real physical world, time cannot stand still. The arrow on the straight line represents the positive and negative characteristics of the matter or particle. As shown in Figure 1-1.

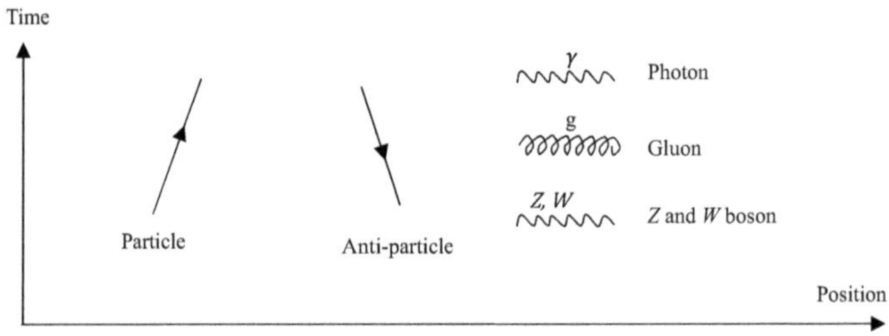

Figure 1.1. Particles in Feynman Diagrams

Ordinary particles in Figure 1-1 are generally represented by solid lines with forward arrows. Reverse arrows can be used to represent antiparticles. Bosons, including photons, gluons, and *Z* or *W* bosons, are generally represented by wavy lines.

The velocity of the particle can be seen from the inclination of the straight line of the particle with the arrow. If the particle's line is vertical, it means the particle is at rest. The larger the angle of inclination toward the position coordinate axis, the greater the speed of the particle. However, compared with the Minkowski coordinates, in order to express the state of the particle more intuitively, the Feynman diagram does not strictly map the tilt angle of the particle to the velocity one-to-one. Therefore, Feynman diagram is actually a qualitative analysis method. Of course, through some conventions, some Feynman diagrams can also be converted into corresponding formulas for calculation.

If you want to represent the interaction between two particles, since there will be a change in position after the particles interact, you can use a point to represent the inflection point. This inflection point usually means that a special space-time location may appear when interacting.

Figure 1-2 shows the interaction of an electron and a positron. In the figure, we can see that there is an obvious inflection point at the moment when the positron and the electron interact, that is, when the virtual photon is transmitted. This inflection point shows a change in the direction of movement of the positron and the electron.

To show the specificity of this inflection point, we can mark this inflection point with a black circle or a hollow circle. In the figure, we use black filled circles to represent positrons, and open circles to represent electrons. Of course, this has something to do with each person's habits, as well as with specific problems. That is to say, when drawing Feynman diagrams, there is actually no very strict convention. After all, this tool is essentially a qualitative tool.

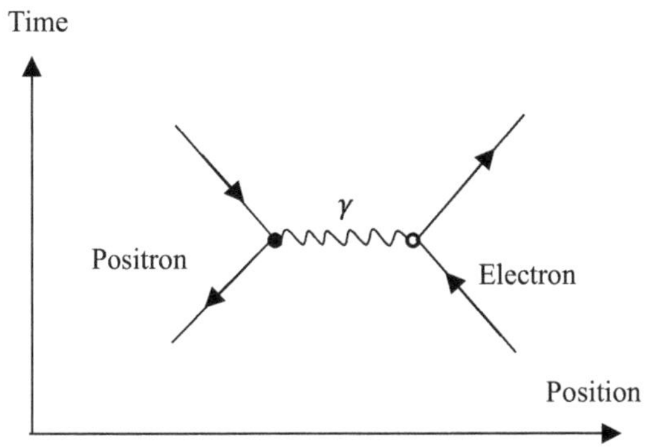

Figure 1.2. Particles in Feynman Diagrams

From the figure we can see that a positron move from left to right positions while an electron move from right to left positions. When the electron and positron are located very close together, the electric fields of the two particles will interact and cause virtual photons to pass between them. The transmission of virtual photons means that electromagnetic interaction occurs.

Electromagnetic interactions due to the transfer of virtual photons will cause changes in various parameters and positions of positrons and electrons. If it is an elastic collision, that is, after the electromagnetic interaction between the two particles, only the energy is exchanged, and no other particles or extra photons are generated, then after the virtual photons are exchanged, the positron and the electron will be separated. As can be seen from Figure 1-2, after the interaction, the positron will move from right to left, and the electron will move from left to right (note that the arrow here does not represent the direction of movement). In addition, we can also see from the figure that since the absolute values of the tilt angles of electron and positron are equal before and after the collision, that is, before and after the electromagnetic interaction by exchanging virtual photons, the velocities of the positron and electron before and after the collision are equal to each other.

After Chapter 2, we will introduce Feynman diagrams to describe interactions in social systems. The advantage of such a diagrammatic approach is that it can well describe the evolution of social systems over time. This is to describe the social system in a dynamic way, which is different from the description of the social system in a static way.

1.1.3 Basic parameters of thermodynamic

For the interaction between two particles, we have been able to use Newtonian mechanics or Maxwell's equations to solve them accurately. However, more complex interactions, involving three particles, become very complex. Even now, for the interaction of more than three particles, we still do not have a general and accurate solution.

For such complex problems, in fact, humans already have a very effective tool to solve the problems involved. This tool is thermodynamics.

The so-called thermodynamics is to study the laws of motion of complex systems from a macroscopic point of view. Although a system is very complex on a microscopic level, involving the motion of billions of billions of molecules, it is completely impossible to accurately solve it by classical mechanics, electrodynamics, quantum mechanics, etc., but on a macroscopic level, we can still through setting certain parameters, and then observe the changing relationship between these parameters to predict the possible movement trend of the entire system.

For thermodynamic system, the three most basic parameters are pressure p, volume V and temperature T. The pressure corresponds to external or internal interactions, and the presence of pressure changes the shape of the thermodynamic system, which is the volume of the thermodynamic system. As a thermodynamic system, of course, one of the most important parameters is the temperature of the system. The temperature of the system is directly related to the heat absorbed or released by the thermodynamic system.

Therefore, as long as we can establish a changing relationship between p, V and T, then we have established the most basic equations of the thermodynamic system. Just like Newton's second law.

In thermodynamic, this formula similar to Newton's second law is called the equation of state of a gas. The general form of the equation of state for an ideal gas is

$$pV = NRT \tag{1-3}$$

where p is the pressure of the gas and V is the volume of the gas. R is a constant and N is the number of matters. T is the absolute temperature.

However, when dealing with thermodynamic system problems, there will be differences in form due to the different parameters involved. If the matter number N is used, the equation can take the form of Equation (1-3). But if we can count the number of molecules directly, we can use the following form

$$pV = nk_B T \tag{1-4}$$

Among them, n is the molecular number contained in an ideal gas, k_B is the Boltzmann constant.

$$k_B = 1.38 \times 10^{-23} J \cdot K^{-1} \tag{1-5}$$

Of course, the Boltzmann constant can be converted to R by multiplying it by the Avogadro constant. Which is

$$R = k_B N_A \tag{1-6}$$

where N_A represents the Avogadro constant. $N_A = 6.02 \times 10^{23}\, mol^{-1}$

Therefore, the equation of state of an ideal gas can also be expressed using the Avogadro constant. which is

$$pV = \frac{n}{N_A} RT \tag{1-7}$$

Of course, in the actual calculation process, you only need to use a form that you are familiar with.

There are three main laws associated with the PVT equation in thermodynamics

The first law is called the Boyle-Mariotte law. This law says that under isothermal conditions, the volume of a gas is inversely proportional to the pressure. The greater the pressure, the smaller the volume, the greater the volume and the lower the pressure.

The second law is called Charlie's Law. In Charlie's Law it states that if the pressure of an ideal gas is constant, an increase in temperature will result in an increase in volume.

The third law is called Gay-Lussac's law. The law states that for a constant volume, pressure and temperature are proportional.

In addition to the three most basic thermodynamic parameters, pressure, volume and temperature, we can use several other parameters when dealing with thermodynamic systems.

One of them is the chemical potential μ. The chemical potential reflects the change of the matter number N or the number of molecules n of the system.

Internal energy *U* and heat *Q* are directly related to temperature. The higher the internal energy, the higher the temperature. Of course, according to the zeroth law of thermodynamic, a system with a higher temperature will always transfer heat to a system with a lower temperature. When a system absorbs heat, its internal energy rises again. Internal energy and heat are related to the energy *E* input and output of the entire system and the environment. Of course, when we solve various physical problems, we are actually using the most basic law of conservation of energy. Therefore, in the thermodynamic system is no exception. By solving the ideal gas equation of state, we can find changes in system energy, internal energy and heat, etc. And then by the law of conservation of energy, we can accurately predict the future change direction of the entire system.

In mathematics, we analyze a curve mainly by taking its derivative. By taking the derivative, we can obtain some very important parameters of this curve. For example, where is the maximum value and minimum value of the curve? where does it have an inflection point? and so on. The change of a thermodynamic system is actually the same as the change of a curve in mathematics. Therefore, when deriving predictions in theory, we also hope to have some very important parameters as criteria to judge what state the system may be in.

In this way we can get some other very important thermodynamic parameters. These parameters include entropy and free energy. Entropy and free energy are used as the criterions to judge which state a thermodynamic system is in. There are many types of free energy that we can use. Including the commonly used Helmholtz free energy *F*, Gibbs free energy *G* and so on.

In order to facilitate social science researchers who are not familiar with thermodynamic to have a clearer concept, Table 1-1 categorizes these parameters.

Table 1.1. Commonly used thermodynamic parameters

Parameters	Symbol	Meaning
Basic	p	Pressure
	V	Volume
	T	Temperature
	μ	Chemical potential
Energy	E	Total energy
	U	Internal energy
	W	Work
	Q	Heat
Status	S	Entropy
	H	Enthalpy
	F	Helmholtz free energy
	G	Gibbs free energy

1.1.4 Equilibrium thermodynamic and non-equilibrium thermodynamic

In Newtonian mechanics, if an object is in equilibrium, the object will be at rest or keep moving at a constant speed. In the thermodynamic system, if all the parameters of a thermodynamic system can be finally balanced, then the thermodynamic system will enter a relatively stable state. In this stable state, various parameters will no longer change. Moreover, the system no longer has energy input and output or the energy input and output of the system are balanced, which will cause the system to consume no additional energy. If we calculate the entropy of the system, we can also find that the entropy of the system will be at the maximum state.

This kind of thermodynamic system that is very stable, has no energy consumption, and has the maximum entropy of the system is called an equilibrium thermodynamic system. Equilibrium thermodynamic systems can be described directly using the equation of state for the gas.

But we can also notice that if the parameters of the system change, such as externally applied pressure, system volume or temperature, then the stability of the system will be destroyed. Changes in one thermodynamic parameter will cause other thermodynamic parameters to change accordingly. The constant changing process of these thermodynamic parameters means that the entire thermodynamic system has begun to appear unstable. In this unstable state, the input and output energy and the entropy of the system may change continuously, and the system will absorb or release additional energy from the environment. For a completely adiabatic thermodynamic system, such a non-equilibrium state means that the work and internal energy of the system will be converted. Such unstable systems are called non-equilibrium thermodynamic systems.

For an equilibrium system, its thermodynamic parameters are relatively simple, and the relationship between various coefficients is also linear. Therefore, equilibrium systems are generally also called linear systems.

But for non-equilibrium systems, the situation becomes more complicated. Nonequilibrium thermodynamic systems can be divided into linear nonequilibrium systems and nonlinear nonequilibrium thermodynamic systems.

Linear non-equilibrium systems are always unstable. That is to say, when the linear non-equilibrium system develops to the final state, it generally enters another equilibrium state. But there may also be some special cases. If there is continuous energy input and output, the non-equilibrium state can also be maintained. For example, the two ends of an iron rod are located at two infinite heat sources with temperature differences, so a temperature gradient can be maintained in the iron rod all the time. The existence of temperature gradients reflects that the thermodynamic system is in a non-equilibrium state.

A nonlinear non-equilibrium system may be a self-organizing system. For example, in an ecosystem, each biological population can maintain a quantitatively stable state. This steady state results from the nonlinear relationship in which different biological populations support and inhibit each other.

Since the system is nonlinear, it is used to reflect the state function of the system, such as entropy or free energy, etc. The change of these parameters is a curved curve. Since it is a curved curve, it means that there may be a maximum or minimum value of entropy on the curve.

If the curve of the entropy function is at a minimum value, it means that the position of this minimum value will be a relatively stable state. Because every time the entropy function deviates from the minimum, the system has a mechanism to pull it back to the minimum. It's like a small ball inside the bowl that will stay firmly at the bottom of the bowl.

However, unlike the equilibrium thermodynamic system where the entropy is at its maximum value, the nonlinear non-equilibrium system needs to consume additional energy to maintain the minimum value of the entropy function. Therefore, a non-equilibrium thermodynamic system can also reach a certain stable state, but it needs a continuous energy input to maintain it. This is what the dissipative structure means. In addition, it is emphasized that this stable state is the opposite of equilibrium thermodynamic, in which entropy is at a minimum value.

1.1.5 Linear approximations for nonlinear nonequilibrium Thermodynamic

Humans already have very effective thermodynamic tools to deal with the problems of equilibrium thermodynamic systems. But for the nonlinear non-equilibrium thermodynamic system, it is very difficult to carry out the calculation. The main reason is that the structure of the nonlinear non-equilibrium thermodynamic system is relatively complex, and it is often difficult to use several thermodynamic parameters, such as pressure, volume and temperature, to reflect the changes in the state of the entire system. Therefore, to solve the problem of nonlinear non-equilibrium thermodynamics, it is usually necessary to solve some very complex nonlinear differential equations. For extremely complex ecosystems, such as human society, the establishment and solution of such nonlinear differential equations becomes more difficult.

However, in physics, there are actually many very effective ways to solve these complex nonlinear problems. These methods include linear approximations, perturbation methods, and more.

For nonlinear non-equilibrium thermodynamic systems, to make a linear approximation, we can first perform a hierarchical partition of these systems. Although the system is complex, the functions and mechanisms involved are relatively simple when entering each level.

Let us first make a hierarchical division of the composition of human society. Because we want to use the thermodynamic method in physics, the hierarchical division we do here will start with elementary particles and continue to the formation of human society. The whole hierarchical relationship is roughly like this: elementary particle → atom → molecule → cell → tissue → organ → individual → social atom or molecule → basic social member → society.

The specific hierarchical structure and classification relationship can be represented by Table 1-2.

Table 1.2. The hierarchy of society

Sociology	Society Basic society members Social atoms or molecules Individuals
Biology	Organs Tissues Cells
Physics	Molecules Atoms Elementary particles

Among the above-mentioned hierarchical structures, elementary particles belong to the simplest level. It can be described directly by the Quantum Field Theories and Schrödinger equation of quantum mechanics. Of course, if it involves the decay of elementary particles, the structure of quarks, etc., complex nonlinear problems will also be involved at this time.

However, the three main particles that affect the formation of atoms and molecules are electrons, neutrons and protons. The atoms and molecules made up of these three particles begin to grow in complexity. Especially at the level of biological macromolecules, due to the different arrangements of base pairs, the complexity of genetic matter DNA in organisms becomes very large. This involves more complex nonlinear problems. Of course, we cannot use quantum mechanics to accurately solve the DNA biomacromolecule. In Table 1-2, we can categorize the three levels of elementary particles, atoms, and molecules into the "Physics" category. Problems at these levels are mainly dealt with by physical methods. Such as quantum mechanics, electrodynamics, Newtonian mechanics, thermodynamics, etc.

When the above-mentioned hierarchical structure enters the level of cells, the complexity brought by quantum mechanics gradually weakens. At this time, we no longer need to use quantum mechanical methods in the study of cells and the tissues and organs formed by cells.

This can see the benefits of hierarchical division of complex systems. This is the ability of hierarchical division to have a kind of encapsulation. It is as if in programming theory, object-oriented programming can encapsulate a complex program (Rentsch, 1982) to form an object. When compiling a new program, this object can be used directly without taking into account the complex internal program encapsulated in this object. External programs only need to use a few important properties of this object.

When studying biology, we do not need to know the principles of quantum mechanics involved in the formation of these DNA biomacromolecules, we only need to understand the coding structure of these DNA biomacromolecules. This means that after hierarchical division of complex systems, at some higher level, we no longer need to care about the very complex computational details involved in the lower levels, including the complex nonlinearities structures involved. This allows us to use a linear approach to solve the higher-level problems. For example, we do not need to understand the principles and calculation methods of quantum mechanics, we can directly use biochemical methods to arbitrarily edit the genetic matter, and then control the growth of genes.

Because the knowledge of dealing with the three levels of cells, tissues and organs mainly involves biology and other knowledge, we therefore classify these three levels into the "biology" category.

At the level of social individuals, we can carry out a new encapsulation. Within this package, each social individual possesses some very important characteristics. Such as the gender of social individuals, the information conveyed, the age and so on. Of course, these characteristics are closely related to the biological structure and the functions of various organs within the social individual. However, after encapsulating each social individual, we do not need to care about these low-level very complex biologically non-linear characteristics. For example, we only know that this social individual has sent a message, and this information has been transmitted to another social individual, resulting in an interaction between the two social individuals. This interaction leads to further corresponding actions between the two social individuals. Even when dealing with broader problems, we don't need to care about how the neural networks of the two social individuals process this information, because the nonlinear process of processing information in such a highly complex neural network of the brain has been completely encapsulated. In this way, at the level of social individuals, we can use linear methods for calculation and processing.

We use this linear method to deal with the interactions between social individuals, and we can find that social individuals can also form structures similar to atoms or molecules. And the atomic or molecular structure formed by this social individual can further interact to form a more complex structure of social thermodynamic. Considering the similarity in some aspects between the structure of social thermodynamic and physical thermodynamic system, we can naturally borrow many laws from physical thermodynamics to deal with some problems in Social Thermodynamics.

Therefore, in the end, we classify the four levels of social individuals, social atoms and molecules, basic social members, and society into the "Sociology" category in Table 1-2. Knowledge of dealing with this category relies mainly on sociologic knowledge.

With the above hierarchical division of the social system, a corresponding solution method can be used for each level of thermodynamic state.

We now have a lot of knowledge about equilibrium thermodynamic systems. With our current knowledge, we know that this equilibrium thermodynamic system is linear. Therefore, at any level, if we can treat the thermodynamic parameters in the social thermodynamic system as the same parameters as the physical thermodynamic system, such as those thermodynamic parameters shown in Table 1-1. We can think that this biological thermodynamic system or social thermodynamic system is also linear.

So in this higher-level thermodynamic system, what we need to do is to find out the social pressure corresponding to the physical pressure, the social volume corresponding to the physical volume, and the social temperature corresponding to the physical temperature. And these thermodynamic parameters should have the same unit, so that all physical parameters can be compared with social parameters, and then use various equations in thermodynamic theory to do further calculation research.

If we determine the corresponding relationship between the social thermodynamic parameters and the physical thermodynamic parameters, we can also determine the conditions for the social thermodynamic system to reach equilibrium or non-equilibrium. The criteria of these important thermodynamic systems have obtained very effective practical tests in physical thermodynamic systems, and they are naturally more objective and effective when applied to social thermodynamic systems.

Although higher-level social thermodynamic system problems can be addressed in a similar way to physical thermodynamics, we can also note differences between them. The lower levels in the physical thermodynamic system, the interactions between molecules, are actually mostly linear interactions. Therefore, the physical thermodynamic system does not consume additional energy when it reaches a state of equilibrium.

However, for the high-level social thermodynamic level, although we can also use the static linear approximation method in the physical thermodynamics to deal with the problem of the social system, the stability of such a social system still needs to rely on the constant input of external energy. It is just that this external energy input maintains the biological and physical functions of the social system, and will not affect the high-level conditions of the social system to achieve linear approximation, nor will it affect the use of the equilibrium thermodynamic method to study the social thermodynamic system.

1.1.6 Van der Waals equation and phase transition

If there is no interaction between gas molecules, the volume of gas molecules can also be ignored, such thermodynamic system is the simplest. It can be directly described by the ideal gas equation of state. If there is interaction between molecules, appropriate corrections to the ideal gas equation of state are required. Such gases with internal interactions are called non-ideal gases.

For non-ideal gases with interaction between molecules, the most successful one is the Van der Waals equation. Van der Waals assumed that there is mutual attraction a between non-ideal gas molecules, and the molecular itself has a volume b. These parameters will have a significant impact on the ideal gas equation of state.

The Van der Waals equation has the form:

$$\left(p+\frac{a}{v^2}\right)\left(v-\frac{b}{n}\right)=k_BT \tag{1-8}$$

among them

$$v=\frac{V}{n}$$

where p is the gas pressure, v is the gas volume occupied by a single molecular (represented by lowercase here), n is the number of molecules, a represents the interaction between the molecules, and b represents the volume of all molecules. T is the temperature and k_B is the Boltzmann constant.

However, we are more accustomed to using V in specific calculations, so by substituting $V = nv$ into formula (1-8), we can get

$$\left(p+\frac{n^2a}{V^2}\right)(V-b)=nk_BT \tag{1-9}$$

Solutions for different states of matter can be obtained using the Van der Waals equations, such as gas, liquid or solid. The ideal gas equation of state cannot obtain these results. If applied to social systems, it can predict what different states a social system will be in in the future.

The solution to Van der Waals equation (1-9) is

$$p = \frac{nk_B T}{V - b} - \frac{n^2 a}{V^2} \tag{1-10}$$

Figure 1-3 shows the curve for one of these temperatures. Since all temperatures on this curve are equal, it is also called an isotherm line.

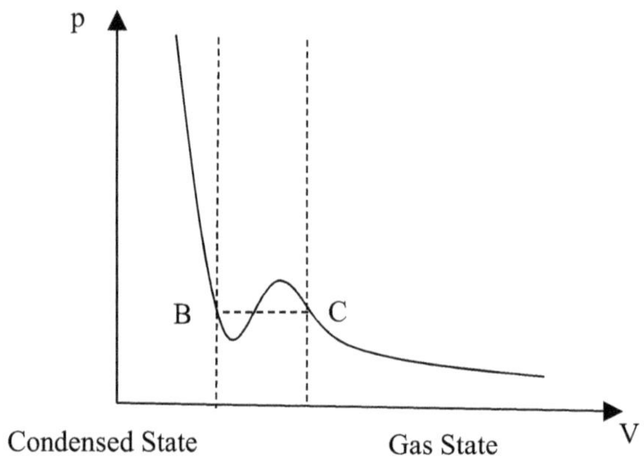

Figure 1.3. A typical solution of van der Waals equation

From the typical solution of the Van der Waals equation shown in Figure 1-3, it can be seen that when the volume of the substance is relatively small, the change in pressure has less effect on the volume of the substance. This is basically the same as the behavior of matter in a liquid or solid state. This state is called the "condensed state" in physics, and corresponds to the position where the volume in the graph is smaller than point B. And if the volume of the substance expands relatively large, the external pressure changes slightly at this time, and the volume of the substance will change greatly. This is basically the same as when the substance is in the "gaseous state", which corresponds to the position where the volume is greater than point C in the figure.

The gaseous state corresponds to an unstable state, because a gaseous state means that the volume is easily changed by changes in pressure. The condensed state is a relatively stable state.

When the volume of the gas is relatively large, this curve can be described more accurately by the equation of state of the ideal gas (1-3). The dotted line BC in the figure is the actual state of the substance when it is in gas-liquid mixing. That is to say, when the volume change is between points B and C, although the volume changes, the pressure and temperature of the substance do not change. This is like when water is boiling, the temperature is always constant at 100 degrees Celsius. Of course, if the air pressure increases, the boiling point of water will also increase.

In the condition of constant temperature, to determine the critical parameters for the phase transition from the gaseous state to the condensed state, we can first take the derivative of V with respect to both sides of Equation (1-9).

We can get

$$\frac{(V-b)dp}{dV}=p-\frac{an^2}{V^2}+\frac{2an^2b}{V^3}=0 \tag{1-11}$$

$$p=\frac{an^2}{V^2}-\frac{2an^2b}{V^3} \tag{1-12}$$

$$\frac{dp}{dV}=-\frac{2an^2}{V^3}+\frac{6an^2b}{V^4}=0 \tag{1-13}$$

This gives

$$V_C=3b \tag{1-14}$$

$$p_C=\frac{an^2}{9b^2}-\frac{2an^2b}{27b^3}=\frac{an^2}{27b^2} \tag{1-15}$$

The critical temperature can then be found by substituting the critical pressure and critical volume into the Van der Waals equation (1-9)

$$T_C = \frac{8an}{27k_B b} \tag{1-16}$$

With these formulas, we can determine the conditions under which the phase transition between the gaseous state and the condensed state occurs by calculating various parameters of the actual gas.

1.1.7 Case study: Critical temperature of water

What we know is that under an atmospheric pressure, water boils at 100 degrees Celsius, that is, 373K, the liquid water and vapor coexists. However, this is not the critical pressure and temperature of water. Because from the van der Waals isotherm, below the critical pressure, even if the pressure and temperature are constant, the volume of water will have a large variation range (Figure 1-3). The significance of the so-called critical parameters refers to the fact that water will change from a gaseous state to a liquid state only when it is lower than this certain volume, pressure and temperature. Otherwise water can only be in a gaseous state.

This temperature can be found using the Van der Waals equation. Of course, considering that water molecules in biological tissues such as the human body account for about 70% of the proportion. This also means that by calculating various parameters of water, the relevant parameters of the social system can be roughly estimated.

We make a transformation of formula (1-15), we can get

$$27ap_C = \frac{a^2n^2}{b^2}$$

so

$$\frac{an}{b} = \sqrt{27ap_c}$$

Then substitute it into formula (1-16), we get

$$T_C = \frac{8}{27k_B}\sqrt{27ap_c}$$

There are three variables in the above formula, Boltzmann constant $k_B = 1.38 \times 10^{-23} J \cdot K^{-1}$. The other two parameters are Van der Waals constant a of water and critical pressure p_c. We now know that the Van der Waals constant of water is

$$a \approx 1.53 \times 10^{-48} Pa \cdot m^6 \qquad (5\text{-}16)$$

This constant is analyzed in detail in Chapter 5. Interested readers can refer to it in advance.

Another parameter is the critical pressure, which is now known as the critical pressure of water

$$p_c = 221.2bar = 2.212 \times 10^7 Pa$$

Note that the International System of Units (SI) is used here, so that we do not need to convert the units in the subsequent calculation process.

This allows us to calculate the critical temperature of water as

$$T_C = \frac{8}{27 \times 1.38 \times 10^{-23}}\sqrt{27 \times 1.53 \times 10^{-48} \times 2.212 \times 10^7} \approx 649(K)$$

This is consistent with current experimental measurements.

Since the atmospheric temperature on the earth's surface is about 293*K* (20°C), water can be converted from gas to liquid on earth. This also means that there is water in a liquid state on the Earth's surface.

The surface temperature of Venus reaches 750*K*, which is significantly higher than the critical temperature of water. This means that on the surface of Venus, it is impossible for water to appear in liquid form. The greenhouse gas effect of water vapor far exceeds that of carbon dioxide, which will make it more difficult for the heat from the surface of Venus to dissipate into space, which further increases the temperature of the surface of Venus.

1.1.8 Relationship between thermodynamic parameters

Although it can be seen from Table 1-1 that there are many parameters in thermodynamics, many of them are related to each other. That is, when a thermodynamic parameter changes, other thermodynamic parameters will also change. This allows us to solve a complex thermodynamic problem with fewer conditions.

The relationship between thermodynamic parameters is based on the first law of thermodynamic, which is the law of conservation of energy.

For a thermodynamic system, there are mainly forms of total energy, internal energy, heat and pressure work. The relationship between internal energy U, heat Q and work W can be expressed by formula (1-17). which is

$$dU = dQ + dW = dQ - pdV \tag{1-17}$$

The differential operator "*d*" is used here, reflecting a microscopic change in a thermodynamic parameter. In the case of macro changes, the following form can be used

$$\Delta U = \Delta Q + \Delta W = \Delta Q - p\Delta V \tag{1-18}$$

However, considering that linear approximations have been made when dealing with the social thermodynamic system problem, the problems involved are basically linear, so without causing confusion, we directly use formula (1-17) to in place of formula (1-18). **That is, the differential operator "*d*" and the macroscopic change symbol "Δ" of the variable are no longer distinguished, unless explicitly stated.**

There is also a relationship between heat and entropy (defined by Clausius):

$$dQ = TdS \tag{1-19}$$

So we can also write the above formula as

$$dQ = TdS = dU + pdV \tag{1-20}$$

There is a direct connection between internal energy and temperature. If the heat capacity of a thermodynamic system is C (constant pressure heat capacity or constant volume heat capacity), then we have

$$dU = CdT \tag{1-21}$$

If the relationship between chemical potential and substance number N is considered, the above relationship of energy change can also increase the energy change μdN due to the change of particle number.

In this way, the law of conservation of energy can be expressed as

$$dU = TdS - pdV + \mu dN \tag{1-22}$$

As for enthalpy H, various free energies, etc., the expression form is detyermined by its definition.

For example,

$$dF = -SdT - pdV + \mu dN \tag{1-23}$$

is derived from the definition of Helmholtz free energy.

1.2 Theories of sociology

1.2.1 The influence of thermodynamic theory on the development of sociology

Sociology is now a very mature discipline, and there are many books on the introductory aspects of sociology. With the increasing development of the Internet, many works on introductory sociology are also open access. This provides great convenience for scholars in different professional fields to study the knowledge of sociology.

Sociology studies the nature, way of life, organization and development direction of human beings and human society. It involves the classification, development, ethnicity, religious belief, gender equality, cultural conflict, and the interdependence of human beings and the natural environment in human society. Sociology is the most basic level of social science. It is strongly influenced by the humanities, social and natural sciences.

The development of sociology can be traced back to the origin of human civilization. The process of how a monarch effectively rules society is described in the earliest human epic "The Epic of Gilgamesh" (Kovacs, 1989). This can be seen as a germ of sociology. Because at this time, human beings have begun to think about the role of various factors in a society through historical events.

However, more than two thousand years ago, during the Spring and Autumn Period and the Warring States Period in ancient China and ancient Greece, with the emergence of ancient philosophers such as Confucius, Plato, and Aristotle, human thinking about society has risen to a more abstract level (Griffiths et al., 2015), and begin to think about the nature of various social phenomena and the interdependence between various social factors.

It was not until the beginning of the nineteenth century that sociology was truly elevated to the level of discipline, when Auguste Comte proposed to use the positivist method to study the laws of society (Comte, 1858). So sociological research has a strict methodology.

After Comte, Karl Marx's sociological conflict theory appeared (Coser, 1957; Collins, 1975). The formation of Marx's philosophical thoughts was mainly influenced by German classical philosophical thoughts such as Kant, Fichte, Schelling, Hegel and Feuerbach (Balibar, 2017; Kouvelakis, 2018). The background of its natural science development is evolution theory, energy conservation and cell theory. Marx expounded the main points of his historical materialism in his book "Critique of Hegel's 'Philosophy of right'" in 1843 (Marx, 1970). This is also an important basis for his philosophical thinking on sociology. In his important work "Capital", which began to be published in 1867, he studied the relations of production (Marx, 2004). This explores the interactions between social individuals at the macro level, providing a theoretical basis for the classification of the six social types. Namely, primitive society, slave society, feudal society, capitalist society, socialist society and communist society.

Durkheim explored social problems from another direction. He proposed that the formation of a good society has a positive impact on the individual. A pathological society leads to various problems for social individuals (Durkheim, 2005, 2016). Therefore, he believes that the structure and function of society can be deduced from the state of social individuals. Durkheim's representative works include Rules of the Sociological Method in 1895 (Durkheim, 2016), Division of Labour in Society in 1893 (Durkheim, 2019).

Another important branch of sociology is Max Weber's Interpretive Sociology (Giddens, 1971). Weber, in his masterpiece, The Protestant Ethic and the Spirit of Capitalism (Weber & Kalberg, 2013), describes sociology as an effort to explain what social action is, and then a causal explanation is given for this social action.

Marx studies changes in social systems caused by interactions between social individuals. Unlike Marx's research, Durkheim and Weber's research is the influence of the whole society on the social individual. It can be seen from this that Marx's research focuses on society, while Durkheim's or Weber's research focuses on social individuals.

Combined with the history of thermodynamics development, we can find that there is a strong correlation between the development of sociology and the history of thermodynamic development. Early Clausius's thermodynamics discussed the problem of thermodynamic from a macroscopic angle. The discovery of the law of conservation of energy, the first law of thermodynamics, led to the emergence of sociology at the macro level, for example, the formation of Marx's dialectical materialist philosophy and social conflict theory and so on. With the more and more profound understanding of the laws at the microscopic level in the thermodynamic system, especially the establishment of quantum mechanics, people can analyze the internal interaction of the thermodynamic system from a lower microscopic level. At this time, the school of "symbolic interaction" in sociology began to appear. By the end of the 1940s, the establishment of Shannon's information theory, the introduction of systems theory and cybernetics, and the establishment of Prigogine's dissipative structure theory in the 1970s made people more inclined to treat a Society as a very complex system. The emergence of sociology of complex ideological systems, including Edgar Morin's Complex Thought, etc., marked the beginning of the theory of modern sociology to penetrate into the micro-level of society. Tables 1–3 contrast the development of thermodynamics with the development of sociology.

Table 1.3. Contrast between thermodynamics development and sociological development

Time	Thermodynamics development	Development of sociology
2000 BC ~ 500 BC	The making of stoneware and bronzeware, etc.	Begin to record history in written form, such as the oldest human epic: The Epic of Gilgamesh
500 BC ~ 1687	Archimedes Statics, Galileo Experimental Physics	Philosophers such as Confucius, Plato, and Aristotle appeared to think about the governance and development of society
1687–1850	Newton's Mathematical Principles of Natural Philosophy Energy conservation law	The establishment of a systematic sociological theory. For example: Comte's Structural-functionalism. His masterpiece "A General View of Positivism" (1848). Marx's Conflict Theory. His masterpiece "Communist Manifesto" (1848)
1851~1905	Clausius: Second Law of Thermodynamics Nernst: Third Law of Thermodynamics	The prosperity of sociological theory. More sociological theories emerged, such as Herbert Spencer. The Study of Sociology (1873) Émile Durkheim. Rules of the Sociological Method in 1895. Max Weber. The Protestant Ethic and the Spirit of Capitalism (1904)
1905 ~1950	Quantum mechanics, Systems Theory, Information Theory, Cybernetics	The emergence of a micro-oriented sociological theory, Symbolic Interactionism. Emerging theories about social individuals and their interactions with society. Such as behaviorism, cognitive theory, Dewey's social psychology.
1950 to present	Prigogine: Theory of Dissipative Structures	Sociological research is developing in the direction of complexity thinking, including Piaget's Constructivism, Edgar Moran's Introduction to Complexity Thinking, etc.

Although from the perspective of the development history of sociology, sociological research seems to have covered the entire macro and micro levels, but if we analyze it from the perspective of research, we can find that almost all sociological research has a research perspective from the human point of view. It also seems that there should be a more macroscopic view of the structure and development of society as a whole.

This macro perspective can be, for example, looking at the life and social system on the entire earth from the globe changing or even the cosmic space, we can find that such a life and social system is actually closely related to the environment of the entire earth, and thermodynamic rules can also be used to explain it.

But this larger perspective is not the same as the God perspective. In various religious theories, God is omnipotent. God can not only dominate the evolution of the entire system in the macroscopic world, but also control the behavior of every social individual in the microscopic world.

1.2.2 Levels of sociological research

The research methods employed in sociology mainly involve qualitative and quantitative research. Qualitative research mainly relies on logical reasoning in language, research on historical documents and archives, interviews, case studies, group discussions, etc. to analyze the laws of society, provide methods for resolving conflicts, and judge the future development trend of society. The quantitative methods mainly include statistical analysis, establishment of mathematical models, controlled experiments, questionnaires, big data analysis and so on.

Quantitative research methods are mainly used in the natural sciences. In sociology, qualitative and quantitative research methods coexist. The two complement each other.

Because the social system is very large, the objects studied by different schools of sociological theory are also very different. Some sociological schools mainly study social systems at the macro level. At the macro level, the social system will be very large, therefore, the object it studies is mainly the structure and overall change trend of the entire social system. Other sociological schools focus on the microscopic level. At the microscopic level, these schools of sociology mainly study the interactions between social individuals. This includes the cognition of social individuals, the interaction between social individuals, and so on. These different schools of sociological theory are like thermodynamics and quantum mechanics in physics. Thermodynamics studies macroscopic phenomena, while quantum mechanics studies microscopic phenomena.

Of course, microscopic laws also affect macroscopic performance, that is, changes in macroscopic parameters. For example, the thermodynamic system formed on the basis of following the statistical laws of classical mechanics meets the requirements of general thermodynamics laws. However, if it is at the microscopic level, the laws of quantum mechanics need to be used to solve the problems, such as the heat capacity of solids, Bose-Einstein condensation, etc.

We can carry out hierarchical division of sociological theories of different schools in order from micro to macro. In this way, it can be seen more clearly which level the objects studied by each sociological school belong to. As shown in Table 1–4.

Table 1.4. The layers to which the different sociological schools belong

Levels	Social theories	Thermodynamic systems
Macro	Social Thermodynamics Conflict theory Structural-functionalism	Societies
	Complex thinking Social constructivism	Basic social members
Microscopic	Role theory Symbolic Interaction Theory	Social atoms and molecules
	Cognitive and Constructivism Behaviorism	Social individuals

As can be seen from Table 1–4, the hierarchy of sociological research is very obvious. At the lowest level of sociological research, it is mainly the research on social individuals. There are already many individual research theories for reference, mainly involving theories such as psychology and cognitive science. Among them, behaviorism theory is the earliest to study the behavior of individuals from a scientific point of view. Behaviorism theory began with the Russian physiological psychologist Pavlov's theory of conditioning (Rescorla, 1967). In the 20th century, Watson's behaviorist theory appeared (Watson, 1913; Watson & Rayner, 1920), and the theory of individual behavior research entered a stage of rapid development.

Cognitive theory (Köhler, 1967; Bandura, 2001) and constructivism (Wadsworth, 1996; Amineh & Asl, 2015) are a level higher than behaviorism theory. Cognitive and constructivism theories touch upon the foundations of a very broad range of physiological, psychological, and cognitive sciences. The theory of cognition and constructivism emphasizes the cognitive process of social individuals, and the knowledge construction of accumulated experience in the continuous cognitive process. This goes beyond the cognitive level of behaviorism's simple stimulus-response. Therefore, cognitive and constructivist theories have a higher level of research on individuals than behaviorism.

If behaviorism and constructivism focus on the study of individual cognitive and developmental processes, then Symbolic Interactionism (Denzin, 2008) and Role theory (Eagly & Wood, 2012) in sociology have risen to a higher level. At this higher level, the two theories study the interactions between individuals. Of course, compared to Social constructivism (Amineh & Asl, 2015) and Complexity thinking theory (Morin, 1992; Montuori, 2008), these theories of interaction between social individuals are still relatively low-level, and more of them involve individuals cognitive activities, etc.

In contrast to complexity theory, etc., Structural-functionalism (Bellah, 2017; Kingsbury & Scanzoni, 2009) has noted the impact of the structure of society as a whole on individuals. Therefore, social changes can be studied through individual changes. This is a relatively macro social study. With the creation of Marx's Conflict theory, Marx has been able to classify different social types and predict the development trend of society from the conflicts and struggles between different classes in the whole society, and then affecting the changes of the whole society. This is a very macro sociological research level.

And if we regard the whole social system as a system the same as the physical thermodynamic system, then we can also study social problems from a more macroscopic level. By applying the laws of various physical thermodynamic systems to social thermodynamics, we can quantify parameters such as social pressure, social temperature, and social space, and express the relationship between these parameters with the formula of thermodynamics. Then, the types of society are divided and what kind of state a society may enter is predicted by calculating and analyzing the results. This is exactly what the book Social Thermodynamics is about. From the perspective of research, if the previous sociological theories mainly studied social laws from the perspective of human beings, Social Thermodynamics can view human society from a more macroscopic perspective of global change. This global change includes not only the transformation of human society, but also the interdependence between human and the environment of the earth, and between human and animal societies. From this more macro perspective, it can be seen that human society is one with the earth-based natural environment in which it lives. This provides a scientific theoretical basis for us to understand and handle the relationship between human survival and development and the earth's environment.

1.3 Combining thermodynamic with sociological theory

1.3.1 Limitations of applying the laws of thermodynamic in society

From the analysis in Section 1.2, it can be seen that the development of sociology is consistent with the development of thermodynamics. As the theory of thermodynamics rises to a new level or new laws are discovered, the research of the whole sociology also rises to a new theoretical level.

But in the absence of tools for theoretical research, the development of thermodynamics mainly affects qualitative research in sociology. For example, through the law of conservation of energy, we know that many laws in the matter world are objective and not dependent on human will. This led directly to the revival of materialist theory. It is against this background that Marxist theory emerged. However, in the past sociological theories, it is still difficult to quantitatively analyze social laws through the law of conservation of energy as in physics. Even if Prigogine proposed nonlinear thermodynamic, dissipative structure theory, etc. (Prigogine, 1978), it directly led to the emergence of sociological theories such as Complex Thought (Morin, 1992; Montuori, 2008). But these sociological theories are still qualitative.

After all, qualitative research cannot be compared with quantitative research. Qualitative research only gives us a general theoretical outline. But to specifically analyze how a society changes from one state to another, what conditions need to be triggered, qualitative analysis seems to be a little helpless.

It can be seen that for any discipline, quantitative analysis is very important and necessary. However, it is very difficult to quantitatively analyze the social system. The fundamental reason for the existence of this huge difficulty is that there is no breakthrough in the understanding of the complexity of the social system.

For example, in history, the lack of a deep understanding of the relationship between human society and the natural environment has led to the emergence of a large number of inefficient heat engines. Although these heat engines can use heat to do work and improve the convenience of human life, they release most of the heat and greenhouse gases into the atmosphere, causing a rapid rise in global temperature and changing the earth environment in which human society lives. The changes in the earth's environment, in turn, cause turmoil and changes in human society.

For another example, when dealing with cognitive problems, although artificial intelligence technology has reached an unprecedented level, it can even realize the application of brain-computer interface to a certain extent. However, the current processing of complex information is based on huge energy consumption. The higher the complexity of the information that needs to be processed, the greater the energy consumption. Just like the current human to process the encrypted algorithm information of Bitcoin, its energy consumption is equivalent to the energy consumption of a medium-sized country. Not to mention the more complex cognitive thinking processes of the brain.

Therefore, it can be predicted that if quantitative analysis is performed according to the computing power of existing artificial intelligence and computer networks, the existing computing power will absolutely not support the computing needs of sociological theories at the micro level.

Even for some macro sociological theories, such as conflict society theory, according to this theory, a society will involve the formation of classes and the struggle between classes, thereby promoting the development of the entire society. In the process of class formation, changes in production relations will be involved. So how to quantitatively analyze production relations is a very complicated problem.

It can be seen that in order to be able to quantify sociological theories and apply the various laws of physical thermodynamic, it is necessary to abstract the interactions between individuals on a more macroscopic scale, and then use a universal concept of interaction replaces the specific content of information exchange. This abstract interaction relationship is then extended to the interaction between social individuals, basic social members, different classes and so on. This abstraction should enable these social interactions to have the same units as those in physics. Such quantified results allow all interactions, whether physical or social, to directly refer to each other in computational formulas. This is also an important content of Social Thermodynamics.

Of course, using qualitative research methods to conduct research also has its advantages. Because qualitative research is used, the generality is very high. For some very complex changes, qualitative research is very effective, although the resulting accuracy will be reduced. This is a reconciling relationship between the complexity of the research object and the complexity of the tools used. The more complex the tools used, the more complex calculations are required. And now you don't have enough tools to do complex calculations, and the result is that you can only use quantitative methods to deal with simpler systems. Conversely, if you adopt qualitative research methods, you can directly use general language to describe the laws of change inherent in complex systems.

1.3.2 Quantification of social parameters

To be able to perform a mathematical analysis of sociology requires the quantification of various parameters related to social issues. There are two options for this quantification process. The first one is a combination of qualitative and quantitative, and only quantifying those parameters that can be directly quantitatively expressed. Such as the number of social population, the amount of food consumption and so on. The advantage of this approach is that it can directly use mathematical methods to quantify parameters related to physical or biomedical laws in sociology, and then use mathematical tools such as mathematical statistics to carry out quantitative analysis. This can provide a quantitative basis for certain sociological conclusions.

For example, the suicide rate of a society can be easily quantified. It can be calculated by calculating the percentage of the number of suicides over a period of time with the total population of the society during this period. Of course, some statistical details are involved, such as error, confidence, etc., and mathematical statistics can be used for more detailed analysis.

After obtaining specific quantitative data in this way, we then compare the results with the results of sociological qualitative analysis to find out what factors have led to an increase in the suicide rate in this society, and then verify a certain sociological theory.

The second one is to build some kind of mathematical model of the parameters, then quantify it according to the requirements of the model, so that it can have exactly the same meaning as the existing physical parameters. In this way, the laws of physics can be transferred into sociology. For example, in the quantification of "demand" in "Economics", by introducing the "marginal effect" model, a person's demand for a certain commodity can be quantified, and then mathematical tools such as the Lagrange multiplier method can be introduced to do the dynamic analysis.

From the perspective of the basic laws of physical thermodynamics, the second method is used to quantify sociology, which mainly involves three important parameters, namely pressure, temperature and volume. These three parameters can be directly measured with the corresponding measurement tools in the physical thermodynamic system. If applied to social systems, although we will refer to corresponding concepts in many literary and artistic works, and even some sociological academic works, in the social sciences, these concepts used to be basically qualitative. This is due to the complexity of the factors involved in social individuals. Even Asch in 1955 conducted a famous experiment in conformity psychology, which reflected the effect of social pressure on changing personal opinions (Asch, 1955). Through this experiment, people can very clearly understand the role of social pressure. However, this social pressure is not quantified. This social pressure is only a qualitative concept, and it cannot be inserted into the calculation formula of physical thermodynamics, because their units or dimensions are different.

Fortunately, if we look at the human groups or other biological groups on the earth from a more macro perspective, such as from the perspective of "global change", we can find that all the effects of pressure, temperature, space, etc. still conform to physical and statistical laws. Because in this perspective of global change, human beings are just one component of a global system that includes the atmosphere, the natural environment, and other living things. At least for the time being, it is not obvious that human society has more advantages than other factors in the overall global system. Of course, if there is a future where the human factor has dominated all other factors in the global system, then we can also study human society from a higher-level perspective, such as the solar system perspective, or more macroscopically, study from the perspective of the Milky Way. Therefore, in the universe, we can always find such a macroscopic perspective, and under this macroscopic perspective, we can make a linear approximation to equate the interaction between humans or biological individuals as similar to electromagnetic or gravitational interactions, in which only the relationship between strong and weak changes.

1. Quantification of social space

Physical volume corresponds to social space. Regarding the understanding of social space, Bourdieu believes that social space is a metaphor (Bourdieu, 1989). Intuitively, social space can be simply understood as the distance between basic society members. So social space is also physically visible. However, in addition to the physical distance that can be measured, social space may also be an individual feeling. For example, individuals in the same class may feel that the exchange of information between them will be effective. If applied to the current "metaverse" of information technology (Heuser, 2003), social space may have a more complex form. This metaverse has abandoned the complete physical space and can be arbitrarily constructed according to the designer's needs by using the network.

The quantification of social space is more straightforward. For example, the land area occupied by a society is multiplied by the average social space height of human society (which can be a constant), so that the size of the social space can be obtained.

However, human society is different from the society composed of animals. Human society also exchanges information with each other. If the channels for information exchange are very smooth, such as the use of technologies such as the Internet, a social individual may feel that his living space has suddenly increased. Such feelings can also be quantified. For example, for a player who is addicted to games, because he has a huge virtual space in online games, he will feel that he can still fully communicate with other people in this society. This is because online games create a huge online "space". This can be seen in the effect of the social space involved in the sudden banning of the player from the game. Because of being suddenly banned from participating in online games, the player will suddenly feel a strong psychological pressure. This is because according to the thermodynamics state equation, the social space in which the player lives is suddenly reduced, and the resulting social pressure will increase rapidly.

The quantification of virtual space can be carried out through a series of physical processes. In Chapter 4 of this book, it is quantified by the change in the number of social individuals that a social individual contacts with other social individuals after the use of information technology. This is similar to in a physical thermodynamic system, if a gas molecule has an increased volume, it means that the gas molecule has more chances to collide with other gas molecules. In this way, by calculating the collision cross section of a gas molecule, the volume possessed by the gas molecule can be estimated. By adding together the volumes possessed by all the gas molecules, the size of the volume occupied by the gas molecules is formed.

If we link the increased opportunities for social individuals to contact other social individuals after using information technology with the social space occupied by a social individual, we can see that different information technologies will have different information spaces. Some highly interactive information technologies, such as the Internet, provide a larger information space for social individuals. While other information technologies, such as traditional postal and telephone systems, provide less information space for social individuals. In the Section 4.3.1, the calculation formula of particle collision cross-section in physics is used to calculate the size of individual information space.

2. Quantification of social pressure

The analysis of social pressure on social individuals is complex. If we observe from the macro perspective of a society with a large population, we can see that these social pressures can actually correspond to the external pressures experienced by the physical thermodynamic system. For example, the social system composed of the Roma people in Europe often makes people wonder that this social system seems to be too fluid. From a thermodynamics point of view, Roma society may be in a liquid-like state. The reason for the relatively high mobility may be due to the relatively large social pressure on the entire Roma society. As a result, this society must constantly flow in order to maintain its own integrity.

Then we can calculate the pressure generated by the society by calculating the energy required by a society when it is flowing, and dividing it by the distance of the society flowing. Since in the calculation process, we use the widely used definition of work done by an external force in physics, the social pressure to be quantified can naturally have exactly the same meaning as the pressure in the physical thermodynamic system. In Section 4.4.2, we used the Black Friday case to define the quantitative parameters of social pressure.

3. Quantification of social temperature

The quantification of social temperature is a bit more complicated. On the one hand, temperature is a very important parameter in physical thermodynamics, which can be measured directly by a thermometer. Temperature, on the other hand, especially when explicitly expressed as social temperature, has very strong literary and emotional implications.

In addition, for social individuals, there is a physiological temperature sensation when socializing with other individuals. For example, studies have shown that a person's body temperature has a strong positive correlation with the individual's feelings in the process of information exchange (Inagaki & Human, 2020). Inagaki et al. found that higher body temperature was associated with stronger social connection. vice versa. Another study by Inagaki et al. also found that individuals felt warmer when reading positive information (Inagaki & Eisenberger, 2013). This reflects that the change of social temperature and the change of physiological temperature felt by such individuals during information exchange have the same neural network mechanism.

From the perspective of the microscopic mechanism of temperature in thermodynamic, the temperature of a thermodynamic system is mainly related to the kinetic energy of the molecules in this thermodynamic system. Molecular kinetic energy directly determines the internal energy of the entire thermodynamic system.

For social systems, the physiological activities of social individuals generate kinetic energy, and therefore also affect the temperature of the society. By calculating the average kinetic energy of each social individual, we can use the calculation formula in thermodynamic statistical physics to calculate a temperature in the social system, which we call the physiological temperature.

However, for human society, the greater energy consumption mainly lies in the information processing of the brain. Since the energy consumed by the brain is proportional to the amount of information processed, we can also directly define another part of the social temperature, the so-called information temperature, as a parameter proportional to the amount of information. Of course, in the calculation of the amount of information, Shannon's theory can already be used to calculate the amount of information such as binary numbers in information theory. In addition, if we express the microscopic molecular motion in a thermodynamic system by Boltzmann statistical entropy, then this kind of information is also directly related to Boltzmann statistical entropy. The calculation and processing of the amount of information in the thermodynamic system is described in detail in Chapter 2 of this book.

In this way, compared with the physical thermodynamic system, the temperature in the social thermodynamic system actually consists of two parts. One is the physiological temperature related to the physiological activities of social individuals, and the other is the so-called information temperature of processing information in the entire social system. At present, there are corresponding theories as the basis to quantify it.

4. Interactions between social individuals

If it involves the interaction between specific individuals, it is of course very complicated, because in addition to general social information, social individuals also transmit emotions, as well as information on the application of various skills and so on. Even if it is to pass general social information, from the perspective of content, it also involves very complex meaning, content, etc. However, we are now analyzing the social thermodynamic system from a more macroscopic scale, so the social individuals we see, or the interactions between social individuals and other basic social members, are actually related to the electromagnetics and gravitational interaction. Because the interaction between these social individuals and basic social members depends on the electromagnetic or gravitational interaction in the sense of physics. Considering that the human body is mainly composed of water molecules, when we quantify the interaction, we can refer to the Van der Waals constant a of the interaction between water molecules, and use this as a basis for estimation.

In this way, when we observe the interaction between individuals in the social thermodynamic system on a macro scale, there is only a distinction between strong and weak. For example, in some cultural backgrounds, the interaction between social individuals will be very strong, but in other cultural backgrounds, the mutual relationship between social individuals and other members will become weaker. In this way, we can introduce different coupling coefficients to the interactions between social individuals in the same way that we introduce different coupling coefficients to various interactions in physics. The greater the coupling coefficient between social individuals or basic social members, the stronger the interaction between these social individuals or basic social members. If this coupling coefficient is weak, it means that the interaction between them is relatively weak, and the organizational structure of the whole society may be looser.

This interaction coupling coefficient can be represented by the constant a in the Van der Waals equation. As for the introduction of information technology, in addition to directly enhancing the interaction, on the other hand, we can actually reflect the changes in the social thermodynamic system through the increase of the social space of individuals. Considering that the constants a and b are very important parameters to determine which state the whole thermodynamic system will enter in the calculation process of the Van der Waals equation, so for the social thermodynamic system, the interaction coupling coefficient and the individual social space size will also directly determine which state the entire society will be in, and ultimately affect the stability of the entire society.

For both human society and animal society, interaction can be divided into two categories, one is gender interaction; the other is social interaction. Both of these two types of interactions are based on the exchange of information, but the effects produced by the two types of interactions will be very different. Their difference can be contrasted with electromagnetic interaction and gravitational interaction. The specific analysis is described in detail in the Chapter 3.

References

Amineh, R. J., & Asl, H. D. (2015). Review of constructivism and social constructivism. *Journal of Social Sciences, Literature and Languages*, 1(1), 9–16.

Asch, S. E. (1955). Opinions and social pressure. *Scientific American*, 193(5), 31–35.

Balibar, E. (2017). *The philosophy of Marx*. Verso Books.

Bandura, A. (2001). Social cognitive theory: An agentic perspective. *Annual Review of Psychology*, 52(1), 1–26.

Bellah, R. N. (2017). Durkheim and history. In *Emile Durkheim* (pp. 3–26). Routledge.

Bourdieu, P. (1989). Social space and symbolic power. *Sociological Theory*, 7(1), 14–25.

Cheng, Z. (2021). *A concise introduction to general relativity*. LAP LAMBERT Academic Publishing.

Collins, R. (1975). *Conflict sociology*. New York: Academic Press.Comte, A. (1858). *The positive philosophy of Auguste Comte*. Blanchard.

Coser, L. A. (1957). Social conflict and the theory of social change. *The British Journal of Sociology*, 8(3), 197–207.

Denzin, N. K. (2008). *Symbolic interactionism and cultural studies: The politics of interpretation*. John Wiley & Sons.

Durkheim, E. (2005). *Suicide: A study in sociology*. Routledge.

Durkheim, E. (2016). The rules of sociological method. In *Social theory re-wired* (pp. 10–15). Routledge.

Durkheim, E. (2019). The division of labor in society. In *Social stratification* (pp. 178–183). Routledge.

Eagly, A. H., & Wood, W. (2012). Social role theory. *Handbook of theories of social psychology*, 2. London: SAGE Publications Ltd.

Einstein, A., Podolsky, B., & Rosen, N. (1935). Can quantum-mechanical description of physical reality be considered complete?. *Physical Review*, 47(10), 777–780.

Giddens, A. (1971). *Capitalism and modern social theory: An analysis of the writings of Marx, Durkheim and Max Weber*. Cambridge University Press.

Griffiths, H., Keirns, N., et al. (2015). Introduction to Sociology 2e. OpenStax. Hoston, Texas. Apr 24, 2015. https://openstax.org/books/introduction-sociology-2e/pages/1-introduction-to-sociology

Heuser, S. (2003). Neal Stephenson's metaspace. In *Virtual geographies* (pp. 171–189). Brill Rodopi.

Inagaki, T. K., & Eisenberger, N. I. (2013). Shared neural mechanisms underlying social warmth and physical warmth. *Psychological Science*, 24(11), 2272–2280.

Inagaki, T. K., & Human, L. J. (2020). Physical and social warmth: Warmer daily body temperature is associated with greater feelings of social connection. *Emotion*, 20(6), 1093.

Kingsbury, N., & Scanzoni, J. (2009). Structural-functionalism. In *Sourcebook of family theories and methods* (pp. 195–221). Boston, MA: Springer.

Köhler, W. (1967). Gestalt psychology. *Psychologische Forschung*, 31(1), XVIII–XXX.

Kouvelakis, S. (2018). *Philosophy and revolution: from Kant to Marx*. Verso Books.

Kovacs, M. G. (1989). *The epic of Gilgamesh*. Stanford University Press.

Marx, K. (1970). *Critique of Hegel's' Philosophy of right'*. Cambridge University Press.

Marx, K. (2004). *Capital: volume I (Vol. 1)*. Penguin UK.

Meyers, R. A. (2003). *Encyclopedia of physical science and technology* (Third Edition) (pp. 617–648). Academic Press.

Montuori, A. (2008). *Edgar Morin's path of complexity*. https://www.academia.edu/213724/Edgar_Morins_Path_of_Complexity. (Retrieved May 5, 2023)

Morin, E. (1992). From the concept of system to the paradigm of complexity. *Journal of Social and Evolutionary Systems*, 15(4), 371–385.

Pathria, R. K., & Beale, P. D. (2011). *Statistical mechanics*. Elsevier Science. The Boulevard, Langford Lane, Kidlington, Oxford, 0X5 1GB, UK: 1–23

Prigogine, I. (1978). Time, structure, and fluctuations. *Science*, 201(4358), 777–785.

Rentsch, T. (1982). Object oriented programming. *ACM Sigplan Notices*, 17(9), 51–57.

Rescorla, R. A. (1967). Pavlovian conditioning and its proper control procedures. *Psychological Review*, 74(1), 71.

Veltman, M. (1994). *Diagrammatica: the path to Feynman diagrams (No. 4)*. Cambridge University Press.

Wadsworth, B. J. (1996). *Piaget's theory of cognitive and affective development: Foundations of constructivism*. Longman Publishing.

Watson, J. B. (1913). Psychology as the behaviorist views it. *Psychological Review*, 20(2), 158.

Watson, J. B., & Rayner, R. (1920). Conditioned emotional reactions. *Journal of Experimental Psychology*, 3(1), 1.

Weber, M., & Kalberg, S. (2013). *The Protestant ethic and the spirit of capitalism*. Routledge.

2

Information processing and the formation of society

2.1 Information and biological interaction

2.1.1 What is information

1. The role of information in cognitive interaction

In cognitive interaction, the psychological interaction involves the transmission of information. In the process of physiological interaction, although the transmission of hormones is mainly involved, the hormones themselves are also involved in the carrying of information. However, for human society, the amount of information transmitted in psychological interaction far exceeds the information transmitted by physiological interaction.

From the classification of social interaction and gender interaction, the process of social interaction mainly involves psychological interaction. Gender interaction is mainly physiological interaction. The information carried in it will naturally be less. However, even in human society gender interaction, the amount of information carried in it is far more than animal gender interaction.

2. Symbol information

There are many different interpretations of information. According to the understanding of people's daily life, information generally refers to various facts about various things, people, environments, etc. (Buckland, 1991). More narrowly, for example, information is data and documents. Even in the United States, information can refer to services about providing phone numbers.

However, in order to be able to reach the requirement that the calculation can be performed, we need to define more precisely what the information is. In this way, information can be represented in mathematical notation.

The first to successfully express information in mathematical language was the American mathematician Shannon. In his monograph published in 1948, he gave the definition of information and the calculation formula of the amount of information (Shannon, 1948).

In Shannon's theory, information is defined as a quantity that reflects the uncertainty of a signal. That is to say, the greater the uncertainty of the received signal, the greater the amount of information carried.

For example, if a coin is tossed, whether it will land front or back, there are two possibilities. So the coin carries information before it hits the ground. After landing, the front and back sides have been determined, and the information about the front and back sides is no longer carried.

So how do you define how much information a coin carries before it hits the ground? Because the coin has both sides, we can directly define the information carried before the coin to land as 2. But this definition will bring a problem, because after the coin has landed, one side has been determined. According to the definition just now, the coin will still carry 1 message after the landing. This is not in line with our daily experience.

In order to solve this problem, Shannon proposed to use logarithm to represent the amount of information. For example, use the logarithm as the base 2 to calculate. In this way, the amount of information carried before the coin falls to the ground is:

$$H = \log_2 2 = 1 \tag{2-1}$$

And the amount of information after the coin landed is

$$H' = \log_2 1 = 0 \qquad (2\text{-}2)$$

This accurately represents the amount of information before and after the coin lands. It can be seen from formula (2-1) that before the coin falls, it carries one piece of information. After landing, the information it carries is 0.

The unit of the amount of information calculated with the base 2 logarithm is "bits". Of course, we can also use the base 10 for the logarithm calculation. It depends on which method is more convenient to calculate for the specific problem. However, when dealing with social problems, using the natural constant e as the base to perform the logarithm operation will bring a lot of convenience in exponential calculation.

We can further generalize this definition, if a thing has N possibilities, the information carried by the thing will be:

$$H = \log_2 N \qquad (2\text{-}3)$$

For example, in a football match, there are three different possibilities for the result, namely "win", "lose", and "draw". Therefore, for the audience, the information of this football match is $H = \log_2 3 = 1.6(bit)$

3. Meaningful information

What is calculated using Shannon's formula is the symbolic information. This does not contain the actual meaning of these character literals. For example, in the same paragraph, the characters and phrases in it are completely shuffled, and the calculated amount of information is equal. But only a few of the sentences are meaningful and understandable.

For example, for the sentence:

Sentence 1: *The key to happiness is accepting one unpleasant reality every day.*

Let's change the position of the phrases in it:

Sentence 2: *The reality to key day accepting unpleasant is every happiness one.*

The notation amount of information in these two sentences is exactly the same. But sentence 1 is meaningful and sentence 2 is meaningless.

For the reader, he can correctly obtain the corresponding information from sentence 1. But for sentence 2, he would find it difficult to understand. Of course, if he is patient enough, he can also get the meaning by adjusting the order of the phrases in sentence 2.

This shows that in a sense, the amount of information of sentence 2 is larger than that of sentence 1. This involves a problem of how to use mathematical formulas to represent meaningful information.

Judging from the feeling of information input into the brain, after sentence 1 enters the brain, it will feel very "orderly". Because it meets the requirements of the basic grammar. This sense of "orderly" corresponds to a lesser sense of "entropy". After sentence 2 enters the neural network of the brain, it gives the impression that it is very chaotic. And this chaotic feeling corresponds to a larger meaning of "entropy".

So here we can try to directly use the "entropy" in physics to express the meaning of a sentence. If the entropy of a sentence is relatively large, it means that the semantic amount of information carried by the sentence is relatively large. On the contrary, if the entropy of a sentence is relatively small, it means that the semantic amount of information carried by the sentence is relatively small.

For the entropy of a sentence, we can use the formula for entropy in statistical physics to calculate it. which is

$$S = k_B lnW \tag{2-4}$$

Here is the natural logarithm *ln*. Where *W* represents number of states. *S* stands for entropy, and k_B stands for Boltzmann constant.

This physical "entropy" corresponds not only to the orderly degree of the information source, but also to the orderly degree of the brain's neural network to process the information. Of course, since it is generally difficult for the brain neural network to completely process the information transmitted by the source, there is also a cognitive coupling coefficient γ for information transmission between the source and the information receiver. The cognitive coupling coefficient can solve how much information is received.

2.1.2 Information and cognition

1. Information and improvement of cognitive ability

According to the simple model of brain neural network (Cheng, 2019), a person's cognitive ability can be divided into two dimensions: cognitive pattern and cognitive depth. The cognitive pattern and cognitive depth are subject to the network capacity and network structure complexity of the neural network.

The network capacity of the brain neural network is directly related to the number of neurons in the neural network. This is consistent with the results in Hopfield's fully interconnected network model in artificial neural networks (Hopfield, 1982).

The complexity of the neural network structure is related to the number of connections between neurons. In a neural network, the more connections between neurons, the more complex the network structure is. The more complex the neural network, the stronger the ability to process complex information (Traub & Woźniakowski, 1990).

However, the physiological growth of the neural network of the brain leads to the rapid growth of its network capacity and network structure complexity, which does not mean that a person's cognitive ability can also increase rapidly. Cognitive ability is also closely related to the knowledge that a person accumulates as he grows up.

This knowledge accumulation is mainly obtained through the continuous deep processing of the information input from the outside world by the neural network of the brain.

From a physics perspective, the input of information constantly brings uncertainty. The neural network of the brain will then adjust its network structure to eliminate the uncertainty of these information. For symbolic information, since the amount of information involved is very large, this needs to be done through enough network storage. That is, a symbol will form a fixed neural network impulse pattern in the neural network. Once this fixed neural network impulse pattern is formed, a cognitive pattern is formed.

In addition to the symbolic information, there is also information about various facts, which is the existence of a basic fact passed to the neural network. Once a fixed neural network impulse pattern is formed in the brain neural network, it will also be remembered as a cognitive pattern.

Neither symbolic information nor information about facts requires deep processing by the brain. Therefore, remembering these symbols and factual information only needs to store data in the brain neural network according to certain clues like data storage, and form the so-called cognitive pattern.

For computer memory, if the speed at which data is stored is fixed, the rate of increase in data in the memory will also be constant. At this time, the corresponding amount of data is also proportional to the time.

However, due to the characteristics of human physiological growth, a person's ability to process information is not constant throughout life. Before a person is 25–30 years old, the neural network of the brain is in the process of physiological growth. At this time, the ability to process information is also growing rapidly. Therefore, the amount of information that can be stored in the neural network of the brain will continue to accelerate with age.

Even by the time the physiological growth of the brain's neural network stops and begins to decline. At this time, the network capacity space is still enough, so there is still enough space for the brain neural network to continue processing information at a high speed, thereby promoting the cognitive pattern to grow at a faster rate.

For cognitive depth, it is related to the interdependence between knowledge. It is usually easier to just learn a simple concept. But this knowledge becomes more complex if a complex concept depends on multiple other concepts. This is like in mathematics if two matrices are multiplied, the problem is relatively simple. But if there are hundreds of matrix multiplications at the same time, the problem becomes very complicated and requires the use of relatively fast computers to complete.

From this we can see that knowledge has a kind of "accumulation". The formation of a knowledge is always closely linked with a series of preceding knowledge. The more prior knowledge is required, the more complex that knowledge becomes. The more complex knowledge needs to consume more resources to process.

Therefore, as long as the neural network structure can provide sufficient computational complexity, the growth of knowledge will be a faster exponential growth until it limited by the network structure complexity.

2. Entropy of information and neural networks

In theory, the semantic information of the sentence can be calculated by calculating the possible number of states of each word in a sentence, and then using the entropy formula (2–4). However, the different experiences and knowledge accumulation of the neural network of the brain lead to different people's understanding of the same sentence will be very different. Those with more relevant knowledge and experience of this sentence will easily understand its meaning. Another person with absolutely no relevant knowledge will find this sentence difficult to understand. This is because for the same information, different individuals have different coupling coefficients with this information.

For the brain neural network, it is mainly composed of a very large number of neurons and more synaptic connections between neurons. If synaptic connections are not considered, and each neuron has only two states of impulse and rest. Then the total number of states owned by N neurons is

$$W = 2^N \tag{2-5}$$

Therefore, the limit value of the entropy of the brain neural network is

$$S_{max} = k_B ln W = N k_B ln2 \approx 0.7 N k_B \tag{2-6}$$

That is to say, the information input from the outside cannot exceed the limit value of the entropy of this brain neural network. This amount of information will be very large. If $N = 10^9$, then $S_{max} = 7 \times 10^8 k_B$. This is a very large capacity, and therefore, the network capacity of the brain is enough to store various complex information.

Although the number of neurons in the human brain is very large, because the impulse of a neuron is always restricted by other neurons, there are not so many possible states of neurons in fact.

Just like the analysis of sentences, if after the first word is determined, subsequent words can appear with a relatively small probability, the semantic information of the entire sentence will decrease rapidly.

If the information is input into the brain neural network, and the pattern that causes the brain neural network impulse already exists in the brain neural network, the probability that the brain neural network predicts the appearance of the next impulse pattern will be greatly improved. This will also help reduce the entropy of the brain's neural network.

This inherent neuronal impulse pattern that already exists in the neural network of the brain reflects the knowledge and experience a person has learned in the environment. The environment that provides this knowledge and experience can be called the "cultural context" in which a person lives.

Of course, if it is unfamiliar information, even if it is inconsistent with the existing impulse patterns stored in the neural network of the brain, as long as the cognitive structure of the brain is complex enough and the computing power is stronger, the neural network of the brain can also generate new information more effectively. The impulse pattern reduces the entropy of the brain neural network generated by the input of this information.

2.1.3 Case study: Bitcoin mining and energy consumption

Bitcoin mining is a very energy-intensive job. There are multiple aspects to the energy consumption of Bitcoin (IEA, 2019). However, the main energy consumption is closely related to the Hash Rate of the Bitcoin network. i.e., how many hashes per second.

The Hash Rate of the Bitcoin network refers to the number of attempts per second that a "miner" needs to mine. For example, a coin has two sides, you only need to try twice or so to be able to turn to the side you want. In this way, the Hash Rate of this coin is 2. Since Bitcoin is a cryptocurrency, it is composed of more binary digits. An exhaustive approach to finding the correct number requires more attempts. In this way, the Hash Rate of Bitcoin will become very large. In addition, the "mining" of Bitcoin will also be artificially limited, that is, the number of Bitcoins that can be "mined" each year is limited. This leads to a lot of computer information processing is mainly wasted on some repetitive work. This will also lead to the higher the Hash Rate of Bitcoin "mining" in the later stage.

Because the Hash Rate of Bitcoin is a very large number, we can also express it in the form of logarithm here.

$$H = \log_{10} \mathcal{H} \tag{2-7}$$

Here $\mathcal{H}$ is used to represent the hash value.

By December 2021, the Hash Rate of Bitcoin has reached 183 *EH/s*. Converted to logarithm, it is

$$H = \log_{10}\left(183 \times 10^{18}\right) \approx 20.26 \tag{2-8}$$

The input of a large amount of information means that a lot of energy needs to be consumed. The energy consumption required for such information processing can sometimes be enormous. It is estimated that the current Bitcoin mining and trading for one year requires electricity equivalent to the annual electricity consumption of some countries.

In 2018, Bitcoin consumed about 0.15 joules of energy per G Hash Rate (De Vries, 2018). Of course this number is constantly changing. With the improvement of the technical level, the energy required to consume a unit of Hash Rate will gradually decrease.

In this way, we can calculate the energy required to consume 183 *EH/s* computing power

$$E = 183 \times 10^{9} \times 0.15 = 2.75 \times 10^{10}\left(J / s\right) \tag{2-9}$$

That is to say, in December 2021, the energy consumed by Bitcoin mining reached a level of 27,500 MW. For comparison, the Three Gorges Dam in China has a power generation of about 22,500 MW. This means that the power consumption of the entire Bitcoin network is more than that of the Three Gorges Dam. This shows that the energy consumption required for information processing is very huge.

It can be compared to the energy consumption of the human brain. If a person consumes $9.4 \times 10^6 J$ per day, the energy consumed per second is 108 joules. A large amount of energy is also consumed in the information processing of the neural network of the brain. Of course, this energy is still very small compared to the energy consumed by Bitcoin mining. It has also been proved that the human brain neural network can reduce external complex information in an appropriate way. This makes it possible to deal with seemingly complex problems with relatively low energy.

2.1.4 Information and interaction

1. Information quantum

In order to clearly see the relationship between information transmission and interaction, here imitating quantum mechanics, each information is defined as a quantum, namely "information quantum".

Information quantum carries all kinds of uncertainty. These uncertainties reflect the size of the entropy of the source itself. However, unlike photons, although information quantum also carries energy, because it can absorb a large amount of energy in the environment, information quantum can be infinitely copied.

This information quantum can be replicated in various forms. There may be an information receiver present, and if she can understand this information quantum, this information quantum will be copied into her brain neural network. Of course, it is also possible to record the information quantum by physical recording, using analog or digital methods, to tape or magnetic disk, semiconductor memory, etc.

In addition, we can further assume that an information quantum carries both energy and state information. This state information can change the state of a neural network in the brain and, of course, the state of an information storage device. We can also use Feynman diagrams to represent this interaction between information quantum and social individuals. The rules for Feynman diagrams are described in detail in Section 1.1.2.

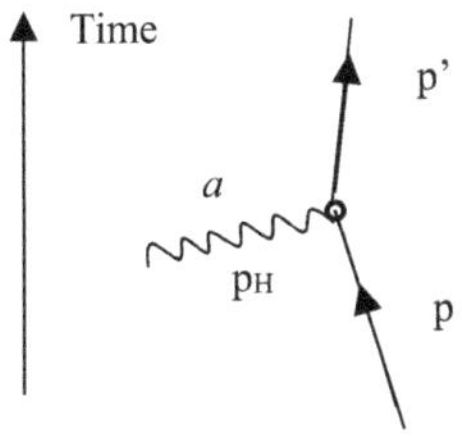

Figure 2.1. Interaction of individual and information

As shown in Figure 2-1, an individual whose initial state is p, receives a message H at a certain moment, and then its final state becomes p'. This way we can list equations like this:

$$p_H + p = p' \tag{2-10}$$

It can also be seen from this that the transmission of information can change the original state of an individual. This is the same as particle collisions in physical systems. When a photon radiates to a particle, it can change the momentum of the particle. This effect is called Compton scattering. This is an electromagnetic interaction. By analogy, we can also regard the transmission of information as a process of changing the state of the individual, which is also regarded as a new way of interaction.

However, in the process of electromagnetic interaction in the physical system, if a particle radiates a photon, the momentum of the particle will also change. This is different from the interaction generated by the information transfer. For the sender of the message, if she doesn't receive any new messages from the outside world, even if she sends out a lot of messages, her state will always remain the same.

2. Information transmission and conservation laws

It will be seen from formula (2–15) that under the condition of constant temperature, the change of energy and the change of information are consistent, namely $dU=TdH$

In this way we can look at information and energy as the same. That is to say, information carries energy. The environment in which the brain is located, however, is an open physical environment with constant energy input. Therefore, for the small system of the neural network of the brain, the laws of conservation of energy and momentum do not apply. In this way, the interaction caused by the information transfer represented by the Feynman diagram cannot be handled by the conservation of energy and the conservation of momentum as in particle physics. This is because a piece of information can be copied indefinitely and then spread to everyone. Take into account that the production of every information requires energy, the transmission of information seems to violate the law of conservation of energy.

The reason for this illusion is that the information transmission process is regarded as a closed system. It seems that energy can be produced indefinitely in this system.

The actual information transmission environment is an open system. For example, in a classroom, a teacher transmits his message to all fifty students at the same time, fifty new messages were created during this process. The generation of these fifty pieces of information naturally consumes fifty times the energy. This energy mainly comes from the nutrients that students absorb, including student breakfast, etc.

Therefore, in the information transmission system, the law of energy conservation still holds. Only the copied information needs continuous external energy input before it can be diffused. This is consistent with Prigogine's Dissipative Structure Theory. That is to say, if the continuous input of external energy is maintained, the entropy of the system can be maintained at a minimum value (Prigogine, 1978).

Another interesting problem is the conservation of momentum. Consider that each individual who receives the information will have further action. This action is accompanied by the movement of the body, the movement of macroscopic and microscopic objects such as the impulses of neurons in the brain. Since the information itself does not carry momentum, it just reflects some uncertainty in the source of the information. Therefore, when this uncertainty is transmitted to the information receiver through different signal transmission methods, such as light, sound and other signals, the information receiver will produce corresponding actions that naturally has momentum.

Although signals such as light and sound also carry a certain momentum, this momentum is still very small compared to the momentum generated by the actions of the information receiver and can be ignored. Therefore, the momentum generated by the information receiver must be conserved within it.

For example, if a pilot receives a message asking him to accelerate, the momentum generated by the acceleration of the aircraft is obtained by burning additional fuel. This maintains momentum conservation within the entire aircraft system. The momentum of radio waves carrying information is completely negligible.

3. The relationship between information quantum and light quantum

Here we compare the information interaction and the electromagnetic interaction, which will help to better understand the difference between the two interactions.

Electromagnetic interaction is carried out by exchanging virtual photons. When a particle emits a virtual photon, the particle will release some of its energy and momentum to the virtual photon. Then after the virtual photon is received by another particle, that particle will absorb the energy and kinetic energy of the virtual photon. This completes the transfer of energy and momentum.

During the whole electromagnetic interaction process, the energy and momentum of the two particles will change. One loses energy and momentum, the other gains energy and momentum. The whole process conforms to the law of energy conservation and momentum conservation.

From the previous analysis, it can be seen that although the process of sending and receiving information still needs to follow various conservation laws, this conservation of energy and momentum mainly occurs within the sender and receiver of information. Information simply reflects the uncertainty of the state of the system. Therefore, the information source that emits the information quantum will not reduce the amount of information it owns due to the emission of the information quantum, that is, the information is sent out, and the entropy of the source's brain neural network will not be changed accordingly. But after the information receiver receives the information quantum, it will change the entropy of his/her own brain neural network. And the change of this entropy is consistent with the received information, and there is a causal relationship.

Second, electromagnetic interaction usually deals with the two-body problem. That is to say, usually two particles collide with each other. If more than three particles collide with each other, it will be a very complex multi-body problem. And if a particle needs to transfer its own energy and momentum to multiple particles, this particle needs to have very high energy and momentum in order to distribute these energies to multiple other particles. In information interaction, a source can easily transmit information to countless information receivers quickly and simply by means of mass media broadcasting, information duplication, etc.

In the electromagnetic interaction, energy and momentum are transferred, and energy and momentum are two parameters that are independent of the material and structure of the particle itself. Information interaction conveys information. Information is a parameter that reflects the entropy of the brain neural network, so the information and the brain neural network are a whole, and it is an important part of the brain neural network. Therefore, the information transfer process is actually a process of changing the entropy of the brain's neural network. From the perspective of cognitive science, the transmission of information is also a process of changing the cognitive structure of the brain.

In this way, we can summarize some important characteristics of information quantum and light quantum (or photons), which are mainly reflected in the following points:

First let's take a look at the difference between the two:

1. Information quantum carries uncertainty, and light quantum carries energy and momentum.
2. There are many transmission methods of information quantum, which can be transmitted using light quantum transmission, sound or even gravitational waves. Light quantum is mainly transmitted in the form of electromagnetic waves.

3. The information quantum can be replicated infinitely, while the light quantum follows the conservation of energy and momentum, and the total amount is limited.
4. The sender can send information without losing energy and can broadcast. However, after photons are sent, and the physical particle loses or obtains energy and momentum.

Although the difference between information quantum and light quantum is very obvious, and it seems that there are few physical laws that can be used when performing information quantum calculations, we can also notice that there are many similarities between information quantum and light quantum. These similarities provide convenience for us to transplant some important laws in physics into information systems. These similarities are mainly manifested in the following points:

1. There are also two kinds of interactions in the exchange of information quantum. Just as the electromagnetic interaction of light quantum is consistent with the strong interaction and weak interaction in particle physics, the interaction of information quantum includes both social interaction and gender interaction. In essence, these two kinds of information interaction are caused by the transmission of information. It's just that the strength of the interaction between the two kinds of information is different.
2. The transmission of light quantum can change the energy and momentum state of a particle. The transfer of information can also lead to a change in the state of an individual. Changes in this state cause changes in the individual's physiological state, which in turn lead to the individual's actions.
3. The interaction caused by light quantum mainly includes two kinds, one is the electromagnetic attraction force, and the other is the electromagnetic repulsion force. The information interaction can also cause attraction and repulsion. When a message causes an individual's "action", this action can include a variety of ways, the most direct one being feedback information to the original message sender. There may also be physical contact

with the original sender of the information. We can also think of physical contact as a form of feedback. In this way, the information feedback of the information receiver can be divided into three situations: attraction, no feedback, and rejection. If there is no feedback, it means that there is only one-way effect, not interaction. There are two ways of attracting and repelling that can cause interaction. These two methods will directly feedback the information to the information sender, which will inevitably lead to a change in the state of the original information sender.

If it is an attractive interaction, it means that the other party accept the information from the source and willing to receive further information. He/She will also feedback information to the source. However, if it is the case of exclusion, it means that the information sent by the source has caused some discomfort to the information receiver, and then the information receiver refuses to receive further information from the source.

2.2 Neural networks and cognition

2.2.1 A simple neural network model

The neural network of the human brain is very complex, which involves both the macroscopic structure, such as the block division of the central nervous network system of the brain, and the fine structure of the neural network, including the connections between neurons and so on.

Due to the complexity of the neural network of the brain, many disciplines have been formed to study the function of the neural network of the brain, including neurobiology, molecular neurobiology, psychology, and cognitive science, etc.

If it is separated from the specific brain neural network, an artificial neural network can be constructed using mathematical models, computer software and hardware. Different from physiology and psychology, artificial neural network uses a mathematical model to construct an abstract model similar to the neural network of the brain, and implements it in software or hardware.

Artificial neural network technology has achieved rapid development in the 21st century when the cognitive function of brain neural network is applied to technology. Especially with the development of deep learning neural networks, the current artificial intelligence technology has been able to obtain very valuable applications in image recognition, natural language translation, automatic driving and other fields.

In addition to using the method of bionics to study the application of the cognitive function of the brain neural network in technology, we can also analyze the cognitive function of the brain through the artificial neural network model. Constructing a computable brain neural network model through artificial neural network enables us to mathematically calculate human cognitive processes. Different from the way of expression in natural language, the derivation of mathematical formulas is highly logical. Once the most basic model is established, various conclusions can be obtained through verifiable mathematical derivation. As long as there are no errors that violate the laws of mathematical operations in the process of mathematical derivation, the conclusions obtained have very strong reliability, unless the most basic model assumptions are wrong. This provides an important basis for ruling out erroneous conclusions and conversely modifying the model.

At present, we already have such an artificial neural network model that can be used to analyze the cognitive function of the brain (Cheng, 2019). This model adopts the simplest fully interconnected neural network model.

Since the actual number of neurons in the brain is very large, it cannot be simulated by computer software or hardware like artificial neural networks. However, because of the large number of neurons in the neural network of the brain, the collective impulse of a large number of neurons can form a corresponding statistical effect. The methods of statistical physics and thermodynamics that we have at present can be well used to analyze it.

2.2.2 Neural network growth and cognition

An important application of the fully interconnected brain artificial neural network model is to study the process of human cognitive growth. Regarding the cognitive growth of children, Swiss psychologist Piaget has summarized four stages of children's cognitive growth through experiments, which correspond to: sensorimotor stage, preoperational stage, concrete operational stage, and formal operational stage (Piaget, 1964). The cognitive growth of adults cannot be studied in the same way as the cognitive growth of children.

Applying artificial neural network models of the brain can solve this problem very well.

First, we build a fully interconnected brain artificial neural network. Then we set the number of neurons in the brain neural network model as $I(t)$. It can be seen that the number of neurons in the brain is a function of time. This means that as a person ages, the number of neurons in the brain is constantly changing.

The changing relationship of the number of neurons in the brain can be described by a Logistic function whose form is as follows:

$$I(t) = \frac{A}{1 + me^{-\alpha t}} \tag{2-11}$$

where α, A, and m are all constants. The constant A reflects the limit of the number of neurons after a person's age reaches adulthood. The constant m represents the growth rate, and the larger the m, the larger the growth rate of the number of neurons per unit time. The constant α represents the increasing time constant. The larger the number of times, the faster the number of neurons will grow.

From the point of view of statistical physics, the more neurons there are, the greater number of states a neural network has, and the more impulse patterns the neural network has. The number of impulse patterns caused by the structure of the neural network itself can be represented by $R(t)$. We can also call this the capacity of the neural network.

The number of neural network impulse patterns will limit the imagination space of the brain neural network. For example, the human brain neural network far exceeds the number of neurons in the animal brain neural network, so the imagination space of the human brain neural network far exceeds the imagination space of animals. We call this imaginary space of the human brain neural network "cognitive pattern", corresponding to the number of things a person's brain can imagine. Here we denote the number of cognitive patterns by $M(t)$.

As can be seen

$$M(t) \leq R(t) \tag{2-12}$$

That is to say, a person's brain cognitive pattern is always less than or equal to the network capacity.

Such an inequality relationship reflects the inconsistency between human cognitive growth and the physiological growth of brain neural networks. During childhood, due to the rapid growth of the brain's neural network, the child's brain's neural network capacity is growing rapidly. However, because a child lacks the necessary cognitive experience, her imagination is relatively lacking, and the corresponding cognitive pattern is not rich enough. Of course, the process of education and learning can effectively improve children's cognitive patterns.

In addition to the number of neurons, there are also interconnections between neurons. The greater the number of interconnections between neurons, the greater the complexity of the neural network. For a fully interconnected artificial neural network, the level of neural network complexity actually directly affects the computing power of the artificial neural network. The stronger the neural network computing power, the higher the complexity of the problems that the neural network can handle. We denote the complexity of this artificial neural network as $C(t)$.

For a person's cognitive process, there is also a level of complexity in solving problems. Usually, the deeper the cognitive process can be carried out, the more complex the problems that can be solved. We call this cognitive ability "cognitive depth", denoted by $D(t)$.

Of course, we have

$$D(t) \leq C(t) \tag{2-13}$$

2.2.3 Asynchrony between physiological growth and cognitive growth

In the relationship between network capacity and cognitive pattern, network complexity and cognitive depth, we can see the incoordination. This dissonance arises from the asynchrony between the physiological growth and cognitive growth of the brain's neural network.

In general, physiological growth peaks faster. That is to say, network capacity and network complexity, which are used to reflect physiological growth indicators, will peak at an early time, and then begin to decline in physiological functions.

But due to the complexity of the knowledge acquired by the brain, it takes time to process this knowledge. The accumulation of knowledge directly affects a person's cognitive ability. This results in a person's cognitive growth process being slower than the physiological growth process. When the physiological growth process reaches its peak, a person's cognitive growth process is far from reaching its peak, his cognitive abilities can still continue to grow.

2.2.4 Division of cognitive development stages based on artificial neural network brain model

The artificial neural network brain model can also be used to analyze the relationship between network capacity, network complexity, cognitive pattern, and cognitive depth.

Since the network capacity and network complexity are different. Children have a certain network capacity as soon as they are born. It also has a preliminary cognitive pattern. This cognitive pattern can meet the most basic stimulus-response needs of young children during their growth.

And network complexity is very small at birth. Although the knowledge acquired by young children at this time is very simple, due to the complexity of the neural network of the brain, the development of cognitive depth of young children is not fast.

By analyzing the relationship between cognitive patterns, network capacity, cognitive depth, and network complexity, we can easily identify the stage characteristics of a person's cognitive growth process.

First of all, in terms of network capacity and network complexity, they reach their peaks in about 25–33 years old, respectively, so a person's cognitive growth process can be divided into two stages, namely, adolescence and adulthood.

Considering the complexity of changes in physiological and cognitive curves during adolescence, it can be divided into four stages:

Stage 1: There is only a small amount of network capacity, but the network complexity is close to 0. At this stage, children only have the basic ability to receive and respond to stimuli. This corresponds to Piaget's sensorimotor stage. This period can also be called infancy. Age is about 0–2 years old.

Stage 2: The network capacity grows rapidly, and the network complexity grows rapidly due to the rapid growth of connections between neurons in the brain. This provides enough space for the cognitive depth of young children to rise, so that children have a certain cognitive depth and can distinguish some more complex things. However, children are not yet able to carry out complex deduction processes at this time. This corresponds to Piaget's preoperational stage. This period can be called early childhood. Age is about 2–7 years old.

Stage 3: Network complexity rises faster at this point, and children begin to be able to relate specific things to each other. Make logical inferences about different specific things. Of course, such reasoning needs to rely on concrete things. This corresponds to Piaget's concrete operational stage. This period can be called childhood. Age is about 7–12 years old.

Stage 4: When the cognitive depth develops to a certain level, the cognitive ability of the neural network of the brain can be separated from the influence of specific things and reason directly through abstract concepts. This corresponds to Piaget's formal operational stage. This is also called adolescence. Age is about 12–25 years old.

This is basically consistent with Piaget's four stages of cognitive development in children. This also proves the rationality of this simple neural network model from one side.

With such a neural network model, we can also divide the cognitive development stages of adults from the interaction of the neural network's physiological and cognitive aspects. This staging of adult cognitive development is based on the fact that, as a physiological system, neural networks undergo a process of growth and decline. The development of cognitive ability depends on the accumulation of knowledge and experience. If the neural network is constantly growing, there will be enough room for the development of cognitive ability. A person's cognitive ability will also continue to increase. If the neural network continues to decline, a person's cognitive ability development space will be restricted by physiological factors, and will not be able to continue to grow.

But physiological growth and cognitive growth are not synchronized. Thus, at a certain stage, cognitive growth coincides with physiological growth. In the next stage, although the physiological growth has reached its peak, it still provides enough space for the growth of cognitive ability, so the cognitive ability can continue to grow rapidly. When it continues, the neural network of the brain begins to decline physiologically. At this time, the growth of cognitive ability gradually reaches the limit of the neural network of the brain, and it cannot continue to grow, and a person's cognitive abilities begin to gradually decline. When a person's physiological and cognitive abilities decline to very low levels, for example, when some cognitive abilities, such as logical reasoning, are lower than those in childhood, and when dealing with life problems, he can only rely on the experience and memory accumulated in the past. At this time, a person has entered the stage of old age.

Following this line of thinking, the process of adult cognitive growth after a person has crossed adolescence can be divided into four stages:

> Stage 1: 25 to 45 years old. At this stage, the physiological development of the brain's neural network peaks and begins to decline. But the physiological system of the brain also provides enough space to meet the needs of cognitive development.

Stage 2: 45 to 60 years old. At this stage, the neural network of the brain begins to show obvious physiological decline. And a person's cognitive abilities are beginning to reach their physiological limits. Therefore, if no special physiological adjustments are made, a person's cognitive ability also begins to peak at this stage.

Stage 3: 60 to 80 years old. At this stage, cognitive ability begins to decline along with the physiological system. However, due to the accumulation of experience, at this stage, many old people can still have enough cognitive ability to surpass many young people.

Stage 4: 80+. At this stage, it officially enters the stage of the elderly. At this stage, due to the strong limitation of physiological factors, a person's cognitive ability is not enough to solve some very complex problems. However, relying on the experience accumulated in the past, the cognitive ability of the elderly is still sufficient to cope with the challenges of various social problems, because cognitive ability is not the only determinant in the process of dealing with the relationship between people.

2.3 Interaction between neural networks

2.3.1 Information exchange between neural networks from a thermodynamic perspective

If information is exchanged between two neural networks, the respective cognitive structures will be changed. This cognitive structure change is analogous to the conduction of temperature in thermodynamics.

For a thermodynamic system in an equilibrium state, the conduction of temperature will cause the entropy of the entire system to gradually increase until the entropy of the system becomes the maximum and the system reaches an equilibrium state. But for a non-equilibrium system, if the continuous input of energy can be maintained, the system will also reach a stable state. However, the entropy of this steady state is at a minimum value. This is the difference between an equilibrium system and a non-equilibrium system (Prigogine, 1978).

Therefore, considering that the operation of the neural network of the brain requires continuous input of energy to maintain a stable state, the brain neural network system can be regarded as a non-equilibrium thermodynamic system. For such a non-equilibrium thermodynamic system, after reaching a steady state, it must be in a state of minimum entropy.

If the information conduction between the neural networks of the brain is regarded as the conduction of this entropy, after the information transmission is over, the two neural networks of the brain will reach a stable state. In this steady state, the thermodynamic system of the two brain neural networks will be in a state of entropy minima.

This is basically the same as the definition of information. In Shannon's information theory, information is defined as a quantity that reflects the uncertainty of an object (Shannon, 1948). The greater the uncertainty of an object, the greater the amount of information carried by the object. This uncertainty corresponds to the "entropy" of the system. Therefore, the process of information reduction is a process of uncertainty reduction, which corresponds to the reduction process of system "entropy".

If information is transmitted between the two brain neural networks, after the information between the two brain neural networks is fully communicated, the uncertainty will be reduced, and the "entropy" of the entire system will naturally be reduced.

If the entropy of thermodynamics is equated with information, then various calculations can be performed here using formulas readily available in thermodynamics. So the Boltzmann statistical entropy is most appropriate here. The information here is represented by H, and the entropy is represented by S. Using the definition formula of entropy in statistical physics, then we have

$$H = S = k_B lnW \tag{2-14}$$

Where, k_B is the Boltzmann constant. W is the total number of microscopic states of the system.

It can be seen from this formula that the greater number of states the system has, the smaller the probability of being in a certain state and the greater the uncertainty. This means that the larger the entropy of the system, the more information it contains.

For the brain neural network system, its temperature is relatively constant, so the information or entropy in formula (2–14) is only related to the number of states of the system and the probability of being in a certain state.

An important formula related to entropy in thermodynamics is the relationship between energy U, entropy S and temperature T, which is

$$dU = TdS = TdH \tag{2-15}$$

In formula (2–15), the relationship between energy U and information H is expressed. It can be seen that the greater the amount of information, the greater the energy consumption required. It also means how much more information there is to process, and how much more energy the brain needs.

For some very complex calculations, equations (2–15) are very obvious. For example, in the calculation process of cryptocurrency, at the beginning, since the amount of information of cryptocurrency is not large, the energy consumption of the computer required is very small. But when the cryptocurrency is about to be fully calculated, the energy consumption is very staggering. Some so-called "miners" who mine Bitcoin will build a hydroelectric power station for their computing equipment to meet the energy needs of information processing.

Since the brain is a very complex nonlinear system, it is also in a non-equilibrium state internally. In this system, although the external amount of information is very large, due to the effect of the nonlinear system of the brain, the amount of information can be effectively reduced through the input of external energy, so that the "entropy" of the neural network of the brain can be maintained at a minimum value. This process is similar to how a computer consumes electricity to process data.

Math further

Of course, in thermodynamics, the second law of thermodynamics states that when a thermodynamic system reaches an equilibrium state, the entropy of the system is always the largest. this means:

$$dS \geq \frac{dQ}{T} \tag{2-16}$$

where Q is heat. If there is no external force to do the work, then

$$dQ = dU \tag{2-17}$$

therefore

$$dS \geq \frac{dU}{T} \tag{2-18}$$

If entropy is regarded as information, then there is also:

$$dH \geq \frac{dU}{T} \tag{2-19}$$

That is, if an information is passed into a closed social thermodynamic system, then no new information enters. We can notice that more new information will be generated after the information is continuously circulated in this social thermodynamic system. This means that even in a closed social thermodynamic system, new information can be continuously generated, resulting in an ever-increasing amount of information. This means that the entropy or information change of a thermodynamic system is always greater than or equal to the change in internal energy divided by the temperature.

But when we study social thermodynamics, we can notice that the social thermodynamic system is a system far from equilibrium. Therefore the production of entropy or information always causes changes in internal energy. Changes in internal energy cause changes in entropy or information. This process is a dynamic one. In particular, the process of dissemination of information is constantly generated, and at the same time, it will be continuously and rapidly processed. In this case, we can always guarantee

$$dH = dS = \frac{dU}{T} \tag{2-20}$$

2.3.2 Cognition and interaction

Another problem with the brain's information processing is the integrity of the information input. It depends on how effectively a person's various sensory organs receive information. Although the neural network of the brain can process various input information, it is limited by the limitations of the sensory organs, resulting in that a lot of information is ultimately incomplete after being input to the neural network of the brain through the sensory organs. For example, visually impaired people have great errors and omissions in the reception of various optical information. Even for normal people, the received information may be lost due to sensory organ fatigue or lack of concentration. In these cases, the processing of information cannot be carried out efficiently.

Therefore, a coupling coefficient γ needs to be introduced here. This coefficient reflects the effectiveness of an independent and complete brain neural network system to receive external information. This coupling coefficient may be dynamic. When a person communicates with different people, the efficiency of communication is different. For example, a person who is very close relation to him can easily communicate. Then the information that the other person sends to him will naturally be easier to deal with. And if he meets a stranger, he may have some difficulty in receiving information from the stranger. This reflects that in the face of different information dissemination objects, the coupling coefficient will be different. In the face of different things, there are also different coupling coefficients. Of course, various parameters in the social system conform to statistical laws, so we can calculate the average value of each parameter. In this way, we can still use the linear approximation method proposed in Section 1.1.5 for analysis.

Here the information coupling coefficient is divided into two parts, one part is called the physiological coupling coefficient. This is a coupling coefficient that is only related to the transmission of sensory organs and hormone signals. The other part is called the psychological coupling coefficient. This depends on factors such as a person's cultural background.

The analysis of the psychological coupling coefficient is more complicated. On the one hand, the psychological coupling coefficient is an information processing method of the neural network of the brain. On the other hand, this approach is a simple information filtering process. Therefore, from the effect point of view, the role played by the physiological coupling coefficient is the same. The introduction of psychological coupling coefficient helps to simplify some problems. For example, although a person's sensory organs are very normal, but due to psychological factors, he refuses to process external information. At this time, external information cannot be input into the neural network of the brain. If it is regarded as an information processing method, it may be necessary to introduce a threshold function to solve this switching problem, and this discontinuous function will bring certain difficulties in mathematical processing. Therefore, taking it as a parameter similar to the physiological coupling coefficient will bring certain convenience in mathematical calculation. However, unlike the physiological coupling coefficient, the psychological coupling coefficient is more uncertain and involves more factors.

The physiological coupling coefficient and the psychological coupling coefficient together constitute the cognitive coupling coefficient. We can express it with the following formula:

$$\gamma = \gamma_s + \gamma_x \tag{2-21}$$

Where γ represents the cognitive coupling coefficient, γ_s represents the physiological coupling coefficient, and γ_x represents the psychological coupling coefficient.

With the parameter of cognitive coupling coefficient, it can be used to process the actual amount of information input to a neural network of the brain.

We multiply the amount of information from the external input by the cognitive coupling coefficient to get the amount of information that can really be processed by the neural network of the brain

$$H = \gamma H_0 \tag{2-22}$$

where H is the amount of information actually processed by the neural network of the brain. And H_0 is the external original amount of information. Combining formula (2-15), we can also calculate that for an external information H_0, the energy that the brain neural network actually needs to consume is

$$dU = \gamma T dH_0 \tag{2-23}$$

This formula also shows that for the same information source, different people need to consume different energy to process this information.

For example, in a conversation between two adults, since the cognitive coupling coefficient between them is relatively large, there will be more information to be processed. In this way, two adult brain neural networks will consume more energy. But for a child who is next to them, because he does not necessarily understand what these adults are talking about, the cognitive coupling coefficient between him and the adults is relatively small. In this way, although there may even be heated arguments between adults, because the neural network of the child's brain does not need to process so much information, his brain consumes relatively less energy for the information discussed by the adult

Taking into account the input of information, it will change the state the network is in. The more information is entered, the greater the change in the state of the network. Therefore, the cognitive coupling coefficient can also be used to reflect the probability of one network transitioning to another network state.

Figure 2-2(a) shows the information transfer between the two networks (N1 and N2), where γ reflects the coupling coefficient between the two networks.

We can also use the form of "Feynman diagram" to represent the state change after the interaction of the two networks. The vertical axis of the Feynman diagram is time, which is continuously increasing. The horizontal axis here is used to represent the state or position of the network. Then the state changes before and after the interaction of the two networks can be expressed as Figure 2-2(a).

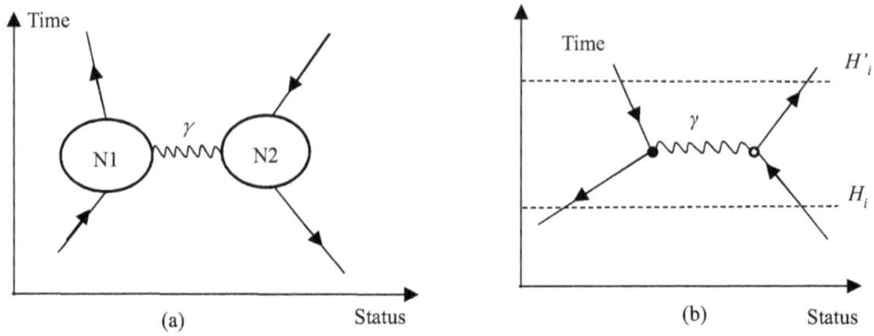

Figure 2.2. The Feynman diagram of the interaction between two networks

Of course, unlike particle physics, the interaction coupling coefficient between networks changes continuously. That is to say, the interaction between the brain neural network of different individuals and other individual brain neural networks has different coupling coefficients. This can be solved using statistical methods.

For easier analysis, we use a circle or black dot to represent a network. In this way, the Feynman diagram of the entire neural network can more clearly see the interaction relationship. As shown in Figure 2-2(b).

The small black circle and reverse arrow in the figure is positive (male), and the small white circle and forward arrow in the middle is negative (female). This can show the complementarity between the two. To emphasize again, in the Feynman diagram, the arrow does not represent the direction of particle movement. Therefore, the arrow directions in Figure 2-2 only represents an attribute of the network.

When calculating, the interaction shown in Figure 2-2(b) can be expressed by the following formula:

$$\Delta H_i = \gamma H_j \tag{2-24}$$

Where $i, j = 1, 2, i \neq j$

Among them, H_i and H_j respectively represent the information that the two neural networks have before they interact. Therefore, the information of the two neural networks after the interaction can be expressed as:

$$H_i' = H_i + \Delta H_i \tag{2-25}$$

To simplify the formula, we use capital letters without subscripts to represent the matrix. which is

$$H = \begin{pmatrix} H_1 \\ H_2 \\ H_3 \\ \cdots \end{pmatrix} \tag{2-26}$$

So the above formula can be expressed as

$$H' = H + \Delta H \tag{2-27}$$

2.3.3 Two types of interactions

For a society with two genders, the cognitive coupling coefficient can be divided into two. One is called the social coupling coefficient. This social coupling coefficient is a common interaction among all individuals in a society. Through the exchange of information, individuals can generate corresponding interactions. Therefore, the social coupling coefficient mainly focuses on psychological coupling. The social coupling coefficient is represented by a. which is

$$a \approx \gamma_x$$

The other is gender coupling coefficient, which mainly exists between two different genders. By exchanging information, more powerful interactions can be formed between different gender members. However, in gender interaction, the information exchanged is very different from that in social interaction. The information exchanged by social interaction is very complex, but the main form is based on various physical signals. The information transmitted by gender interaction is relatively simple, mainly hormone information. However, the strength of the information conveyed by gender interaction far exceeds that of social interaction. The difference between the two can be compared with the gravitational interaction and the electromagnetic interaction. Therefore, gender coupling coefficient mainly corresponds to physiological coupling. Here, use g to represent gender coupling coefficient. That is

$$g \approx \gamma_s$$

In order to distinguish these two kinds of interactions, the gender interaction is represented by a bold straight line, and the social interaction is represented by a wavy line. As shown in Figure 2-3(a).

The Feynman diagram shown in Figures 2-3(b) shows that two social individuals with different genders are combined through gender interaction. Motherhood then produces offspring of two different genders, and interact with them through social interaction. In this way, four of the individuals form a relationship of family members.

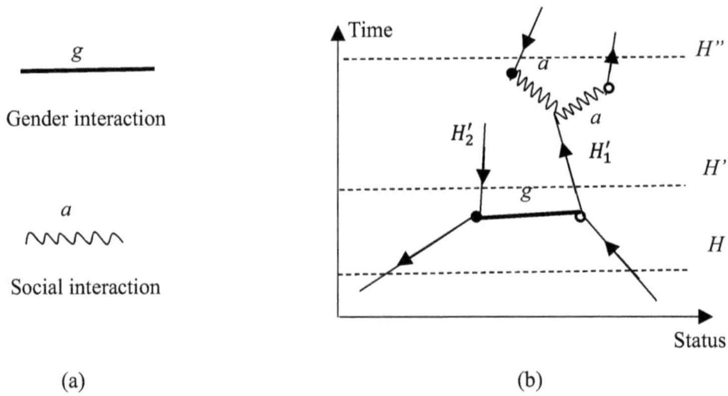

Figure 2.3. The Feynman diagram of gender and social interactions between family members

The calculation formula corresponding to Figure 2-3(b) is

$$\Delta H_i = gH_j \tag{2-28}$$

$$H' = H + \Delta H \tag{2-29}$$

$$H'' = aH_1' = a(H_1 + \Delta H_1) \tag{2-30}$$

It can be seen from the above formula that H'' is only coupled with H_1', so the information obtained by H'' all comes from the information of H_1' coupling.

Combined with the previous formula (2–23) of the relationship between information and energy, we can notice that the change of information is consistent with the change of energy. That is to say, if a family has sufficient energy supply, such as abundant food, the children of this family will receive more information.

Of course, given the complexity of biological systems, human infants not only communicate closely with their mothers, but also communicate with the environment. However, it is believed that in the early childhood, the information exchange with the environment is still relatively weak. If these exchanges of information are to be taken into account, the "perturbation" method often used in physics can be used to deal with them.

2.3.4 Case study: Use Feynman diagram to represent the spread of virus

Viruses do not have gender, so they only need an arrow in one direction. There is no gender interaction between viruses, so the gender coupling coefficient is 0. However, there is information exchange between viruses, so there is a social coupling coefficient between viruses.

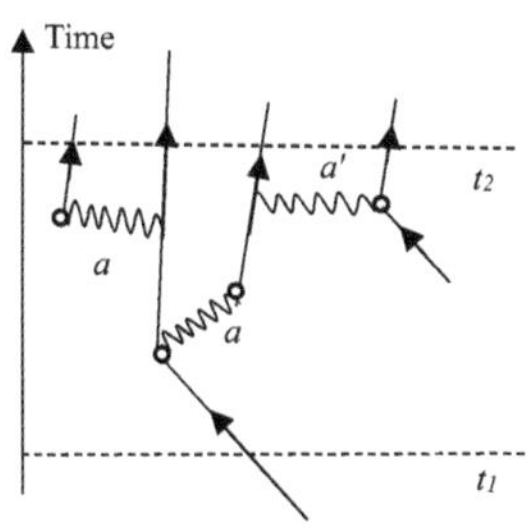

Figure 2.4. The Feynman diagram of the interactions among virus

Figure 2-4 shows the interaction between viruses. In addition to the gene information transmission caused by the replication of the virus, we can assume that there is a relatively weak interaction between the virus and the other virus, such as a' in the figure. Only in this way can the aggregation of the virus be completed. This interaction may also be mutually inhibitory. For example, the mutual inhibition between influenza virus and common cold virus (Nickbakhsh et al., 2019).

This weak interaction between viruses may originate from the physical sense, that is, the kinetic energy generated by the collision of viruses in the process of spreading. There is also the ability to recognize genes. When a virus finds an object that can invade, it will pass the information to other viruses, so as to carry out a centralized attack on some organisms.

As can be seen from Figure 2-4, since viruses can be continuously generated through interaction, the number of viruses also shows a trend of rapid growth. In the figure, there is one virus at the initial time t_1, and after a period of interaction, four viruses have appeared at the time t_2.

2.4 Cognition and the formation of society

2.4.1 A society based on the same cognition

If the cultural background of a society is the same, the cognition of the members of this society is basically the same. This means that the social coupling coefficients between individuals in this society are approximately equal. Of course, considering that the relationship between family members is closer, the social coupling coefficient within family members will be larger. Outside the family, the interaction between all social members has basically the same social coupling coefficient a. In addition, in this society, the gender coupling coefficient g between different genders is also approximately equal. In this way, for this social system, we can use simpler statistical methods to deal with it.

The first is that the two genders are first combined through the gender coupling coefficient. In some social formations, this can be directly called "social molecular". Because it is the basic unit of some specific social formations. Of course, in other social forms, the gender relationship coupled through gender interaction is not enough to constitute the most basic unit of society. It is also necessary to form a family according to the Feynman diagram of Figures 2-5. Because family members have the same background and the same growth process, the social coupling coefficient among family members is stronger than the interaction with other social members outside the family. The formation of a family may also be referred to as the formation of a "family atom". The family atom can be used as the basic unit of a certain social formation.

Compared with the interaction between gender molecules, the interaction between family atoms is more complicated. If the social interaction between family atoms is formed only by exchanging information, the interaction outside the family is relatively weak. Therefore, the corresponding social coupling coefficient will be smaller. But if two family atoms, each of which contributes a social individual with different gender, and combines with each other, at this time, a form similar to the covalent bond in the physical world is formed. This covalent bond is also a form of social interaction. Because this covalent bond involves gender interaction, the interaction between family atoms can also be very strong. In addition to the form of covalent bond, of course the interaction between family atoms can also form a form similar to the ionic bond in the physical world.

Here we use g to denote gender coupling coefficient and a to denote social coupling coefficient outside the family. And we use a' represents the social coupling coefficient within the family. In this way, the Feynman diagram of the whole society can be represented in the form of Figure 2-5.

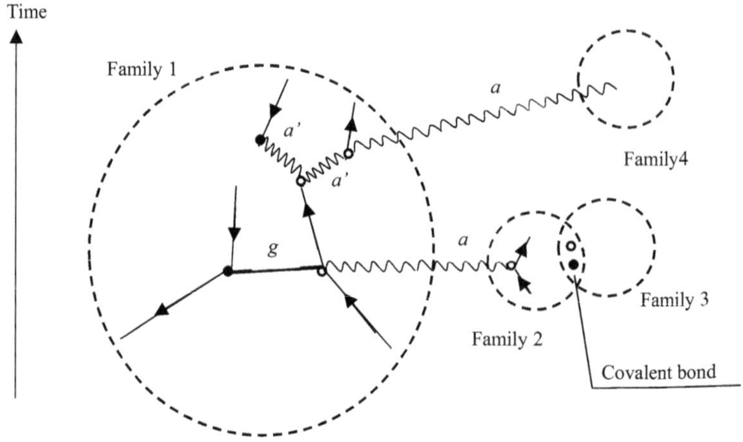

Figure 2.5. The Feynman diagram of the relationship between family and society

If we analyze it from a more macro level. In some social types, the combination of two social individuals of different genders can be seen as a gender molecule. In other types of societies, the union of the two genders is incomplete and further family formation is required, and then each family can be regarded as an "atom". The interaction between these molecules or atoms constitutes a social system. If there is no interaction between gender molecules or family atoms, the social system is analogous to an ideal gas in thermodynamics, in which there is no interaction between gas molecules. This can be calculated using the fundamental equation of state for an ideal gas. It can be calculated using the equation of state for an ideal gas. Although this ideal gas-type society does not exist, its calculations can give us a less accurate result for solving complex problems.

2.4.2 Three different social system structures

The diversity of human society determines that in different societies, the composition of basic society members is very different. In some societies, two different genders can be combined with each other to form an independent basic society member, and then interact with other social individuals to form a complete society. In some societies, the combination of different genders is not complete enough. It is also necessary to have children to form a relatively large "family atom", so that a basic society member can be formed. The society composed of these two different basic society members is naturally very different. This also leads to different types of societies, whose thermodynamic properties will vary widely.

Here, human society can be divided into three types of societies according to the basic society members that constitute the society, namely:

1. Family-oriented society

In this family-oriented society, the family atom is the most basic unit of the social type. Usually the social interaction within the family is much larger than the social interaction outside the family due to kinship.

In addition to information exchange, family atoms can also be linked together by means of "covalent bond", "ionic bond", etc., so the social interaction of family-oriented society is more complicated. Besides psychological interaction, it also involves physiological interaction, so the interaction between family atoms is also very strong.

This family-oriented society mainly refers to those societies with more traditional culture. In these societies, the concept of family is very important. The bonds between family members are very strong. In ancient Chinese Confucianism, "Self-realization as a communal act presupposes a personal commitment for harmonizing the family, governing the state, and bringing peace to the world" (Weiming, 2017), which regards the relationship between family members as the whole the most important element in the social structure.

2. Social-oriented society

In this type of society, its constituent units are social individuals and gender molecules. A social individual refers to a single person. The gender molecular contains two different gender social individuals. In terms of social interaction, the interactions between social individuals and between social individuals and gender molecules are basically similar, and the main forms are psychological interactions.

Because there is no family atom, or even if there is, the social interaction within the family is very weak, so social-oriented society usually downplays gender interaction. The protection of private property is also not as strong as in social-oriented society. Especially with the development of modern society, some social-oriented societies have begun to have a tendency to completely eliminate private ownership. A typical example is the "Paris Commune" that appeared in France in the nineteenth century (Bakunin & Walter, 1971). In addition, Soviet society also belongs to this type (Von Mises, 1981). In such a social structure, the intensity of information interaction between social members is much greater than that between family members.

3. Balanced society

In this type of society, social individuals and gender molecules and family atoms are included. Three type of basic society members coexist in a balanced society. Mutual respect for different cultures forms a multicultural and even multiracial society. And this multicultural society is also a goal that many societies hope to achieve (Jupp, 1997), and has gradually become the mainstream of social types. This is mainly due to the development of Internet technology, which has led to more opportunities for family members to communicate with the outside world. As teens grow, they interact even more closely with other members of society as a whole than with family members.

Considering the complexity of the composition of basic social members, compared with family-oriented society and social-oriented society, the internal structure of balanced society is more complex, involving many nonlinear and non-equilibrium social thermodynamic problems. This may require a dedicated monograph to discuss it. This book does not intend to deal too much with the thermodynamics of nonlinear nonequilibrium stations.

2.4.3 Interaction between different societies

Although we have not yet quantified specific social thermodynamic parameters, we can still do some qualitative analysis of the interactions between societies.

If there is also interaction between two societies of different types of structures, it may lead to some important changes within the two societies. It is like contacting two gases of different temperatures and concentrations.

This can be divided into two cases of mutual integration and mutual independence.

In the case of mutual integration, if the strengths of the two societies are symmetrical, it is still difficult for the two societies to integrate together. Such as the Soviet Union and the United States during the Cold War. The Soviet Union is a social-oriented society, and the United States is a balanced society. The end result of historical development was the disintegration of the Soviet Union.

And if the power between the two societies is asymmetric. That is, one of the society has more members and the social temperature is higher. If the two societies can still integrate with each other, usually the weaker society is easier to integrate into the stronger society. For example, the Tang Dynasty in China was very powerful, and the surrounding ethnic minorities were constantly integrating into the culture of the Tang Dynasty. During the Anshi Rebellion, these ethnic minorities could also be distinguished from the mainstream ethnic groups of the Tang Dynasty. But in modern times, many ethnic groups of that time have been unrecognizable (Lindong, 2008).

Of course, social space does not only refer to the size of the territory, but also to factors such as culture. Therefore, the integration of a weaker society into a stronger society will itself have an impact on the mainstream society. This may lead to resistance from the mainstream social culture. As a result, it appears that they are all in the same country, but the cultural performance of the two ethnic groups is very different. The weaker societies that have integrated in can still maintain a certain degree of independence for a long time. Such as European Romani (also called Roma, Gypsies) (McGarry, 2014), Jews (Gruber, 2002) and so on. This situation has to do with the interaction between two independent societies.

Since there are many types of societies, interactions between independent societies are also complex. Here we focus on the interaction between a balanced society and a family-oriented society.

Figure 2-6 shows two balanced society and family-oriented society in contact. In order to facilitate the distinction, according to the difference in the average number of individuals included in the basic society members, the basic society members of the family-oriented society are represented by large circles. The basic society members of the balanced society are represented by small circles.

There is a border between the two societies. Of course, this border can directly refer to the national border. It can also be the boundaries of a more abstract culture. The so-called cultural boundary refers to those cultures in which the values of both societies can be universally recognized.

As can be seen from the figure, the basic society members in a balanced society include a large number of social individuals, so there are many individuals who communicate with each other. The basic society members in family-oriented society are family atoms, and the number is relatively small. For example, a family-oriented society has one billion people, and each family has at least 10 people. In this way, the basic society members of this society only have 100 million family atoms. Another balanced society has a population of 300 million, and various basic society members have an average of 2–3 individuals. Therefore, the basic society members of this balanced society still have 100–200 million members, which may be more than the family-oriented society next to them.

If the boundaries between these two societies are very strict. That is, there are strong measures to prevent the mutual flow of basic society members. It is difficult to integrate the two societies together.

If the boundary is revoked or the restrictions on the boundary are not so strict, there will be a process of mutual penetration of some basic society members between the two societies near the interface. That is, some members of the family-oriented society will enter the vicinity of the interface of the balanced society. And another part of the members of the balanced society will also enter the family-oriented society. This interpenetration effect can be confirmed by immigration between traditional family-oriented societies such as Japan and Korea and balanced societies such as the United States. Except for some special events that led to a large number of South Koreans entering the United States (such as the annexation of South Korea by Japan, the Korean War, etc.), the growth rate of South Koreans in the United States during peacetime was not fast (O'Connor & Batalova, 2019). Even between 2010 and 2017, the number of South Korean immigrants declined slightly. In contrast, fewer Japanese immigrated to the United States.

However, for a society with the same cultural background, the number of Koreans entering China has increased significantly in recent years, and the number of Koreans in China has also significantly exceeded the number of Koreans in the United States. Although China's restrictions on foreign populations are far stricter than those in the United States.

The interpenetration of basic society members between two different types of societies can be shown in Figure 2-6.

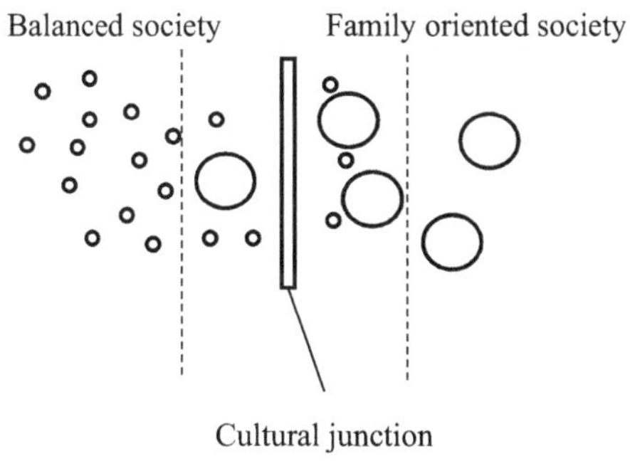

Figure 2.6. Interactions between balanced society and family-oriented society

Figure 2-6 shows a structure similar to a semiconductor PN junction, the boundaries of which are referred to here as "cultural junctions". At the interface of two different types of social contact, due to the mutual penetration of the basic society members of the two societies, a "mixed society" area can be formed on the interface. Two types of basic society members are included in this area at the same time. However, this hybrid society is not stable, because there are many difficulties in the communication between two different basic society members, resulting in a relatively weak interaction coupling coefficient. Therefore, without the continuous input of foreign basic society members, this area will eventually develop into a family-oriented society or a balanced society. It is also possible that the entire society re-splits into two different types of social gathering areas. Therefore, the hybrid society that appears near the interface in Figure 2-6 mainly depends on the continuous input of basic society members on both sides.

The opposing basic society members located near the cultural junction may also continue to spread to the interior. For example, members of a family-oriented society try to penetrate deeply into a balanced society, but at this time, they can find that it will become more and more difficult to spread. After all, the more it spreads into the balanced society, it means that there are fewer and fewer family atoms that can fully interact. At this time, it means that members of family-oriented society either change their cultural beliefs and become members of a balanced society, or return to their family-oriented society.

The same thing happens for a balanced society.

After a certain period of diffusion, the concentration of basic society members in the vicinity of the entire interface will stabilize, and the "cultural junction" will be formed.

2.4.4 Case study: Cultural conflict in schools

In her book "Other people's children: Cultural conflict in the classroom" (Delpit, 2006), Lisa Delpit tells a story about Cathy, a little Asian American girl. This story can be used to analyze the impact that different cultural background conditions may have on the way teachers teach.

In the impression of many Americans, Asians seem to be smart enough to perform well in any academic setting. This has led to the overestimation of the abilities of many Asian children. Lisa Delpit recalled seeing such a scene in the past. Little Cathy is a 5-year-old Asian girl who is also very well-behaved at school. In a student activity class arranged by the teacher, the teacher asked the children to place some objects next to the numbers printed on the cloth. Little Cathy does everything very seriously as usual. During the whole process, Cathy was also very quiet and never disturbed other children. Because Cathy seemed to be doing well, the teacher didn't put more energy on her, and didn't give her any further instructions on how to do it. At this time, the teacher focused her energy on the children next to her who took the initiative to ask the teacher questions or the children who the teacher found difficult to complete the task. Throughout the teaching process, little Cathy is doing this task alone. After finishing, Cathy also packed all the teaching aids neatly. But the careful Delpit found that the result of little Cathy was wrong. She did not place the designated object next to the correct number on the cloth. It can be seen from this that Cathy did not understand the teaching goal of this class. Still, the teachers gave Cathy very high marks, thinking that, like other Asian kids, Cathy was one of the smartest Asian girls they had ever met.

Obviously, if little Cathy was in kindergarten in China, she wouldn't have this problem. Although Cathy looks good-natured, the Chinese teacher can see the problem. Cathy will also conduct in-depth counseling in a timely manner, telling Cathy how to solve these problems.

Here we analyze the problems encountered by little Cathy from the perspective of Social Thermodynamics.

First little Cathy grew up in an Asian family. Asians, especially East Asian societies, belong to family-oriented societies. In such a family-oriented society, the intensity of interaction between members of the family related to information transfer is much greater than the intensity of interaction between members of the external society. As a result, Cathy was definitely more influenced by her family than at school. The values in traditional Asian families: obeying discipline, listening to teachers, not breaking the order, etc., were instilled by parents after birth, and formed Cathy's daily habits. That's why little Cathy appeared as a well-behaved child in kindergarten.

Second, let's look at the cultural background of kindergarten teachers. Cathy's teachers grew up in the environment of a balanced society in the West, so the teacher's value is to have sufficient information exchange with other members of the society, which can meet the requirements of forming stronger relationships between members of the society. As individuals from two different cultural backgrounds, the differences between the teacher and Cathy are very large. This very large difference is enough to create a very serious conflict. But the teacher, as a strong individual in the classroom, has authority. Therefore, in daily teaching, teachers will instinctively express their own values and transmit them to Cathy's brain as information. As a biological individual with a self-organizing function, although little Cathy may not be able to complete the tasks assigned by the teacher, she has her own way to avoid being hurt by the teacher's completely different values.

Again, let's look at the information transfer between the teacher and little Cathy. The basis of this information exchange is that the entropy transmitted to the teacher's brain is reduced. Since little Cathy is very well behaved and disciplined, the information passed to the teacher is naturally that Cathy's task has been completed well, so there is no more information passed from Cathy. On the contrary, for those students who often ask questions or have a poor foundation in the past, and have left a deep impression on the teacher, the teacher will feel that further tutoring is needed, so as to reduce the entropy of the teacher's neural network. This is one reason why little Cathy is often overlooked.

In a family-oriented society, Cathy has a very close relationship with family members, especially her mother, and of course the interaction with other family members is also very strong. In addition, there is interaction between Cathy and the teacher. However, it can be clearly seen that the Cathy family cannot form a solid social interaction relationship such as "covalent bond" or "ionic bond" with other external basic society members in the balanced society. Therefore, the interaction between Cathy and family members is far greater than that with other basic society members, such as interaction with teachers. That is $a' \gg a$.

In this way, the total information received by Cathy from all aspects is

$$H_{Cathy} = aH_{Teachers} + a'H_{Family} \tag{2-31}$$

Of course, if the information $H_{Teachers}$ sent by the teacher is big enough, the interaction between the teacher and Cathy can still have a great impact on Cathy's future growth. But unfortunately, in this example of Delpit, the teacher ignored the instruction of Cathy because Cathy looked too good, which means that in this case, $H_{Teachers}$ is also very small, i.e., $H_{Teachers} \ll H_{Family}$, then the formula (2–31) can be approximated as

$$H_{Cathy} \approx a' H_{Family} \quad (2\text{-}32)$$

This can also be seen why many Asians have difficulty integrating into mainstream Western society even if they grew up in the West.

2.4.5 Case study: Vietnam war

The Vietnam War was a protracted war between the US-supported South Vietnam regime and North Vietnam (Moyar, 2006). The entire war lasted 20 years. If we analyze it from the perspective of Social Thermodynamics. The US belongs to a balanced society, while the whole of Vietnam, including South Vietnam and North Vietnam, belongs to a family-oriented society.

In South Vietnam, there has been a fusion of the balanced society culture represented by the United States and the family-oriented society culture represented by the Republic of Vietnam and South Korea. But in South Vietnam, the balanced society culture in the United States belongs to a minority culture, so the integration process takes a certain amount of time. The two decades of the Vietnam War have seen little change in the culture of the South Vietnam family-oriented society. Instead, a relatively independent so-called "Western culture" was formed within South Vietnam. and subject to a certain degree of resistance from mainstream culture.

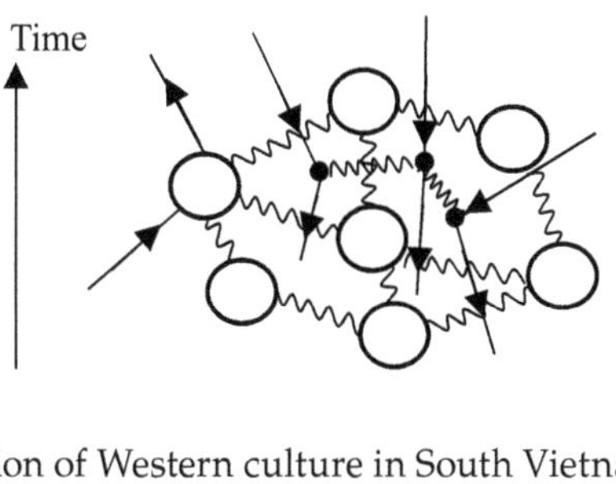

Figure 2.7. The penetration of Western culture in South Vietnam

Figure 2-7 shows the independent development of Western culture within South Vietnam. It can be seen that in a relatively large Asian culture representing a family-oriented society, the Western culture representing a balanced society is very weak. And the interaction between Western culture and Asian culture is very weak. This creates a relationship between two separate cultures. At most, an unstable hybrid society can only be formed at the interface of the two cultures. Moreover, it requires a steady stream of input from basic society members of the two societies to the interface position. But it turned out that the input of the US military was unsustainable. So this cultural junction will eventually disappear.

The difficulty of this fusion between Eastern and Western cultures can also be used to explain the war in Afghanistan that ends in 2021. It also took 20 years from the U.S. invasion of Afghanistan in 2001 to the U.S. withdrawal from Afghanistan in 2021. During these 20 years, there is still no effective fusion between Western culture and traditional Afghan culture.

The fundamental reason why a balanced society is difficult to integrate into a family-oriented society is that the social coupling coefficient of members within a family in a family-oriented society is much greater than that of other members outside the family. And there is a strong social interaction based on covalent bond and ionic bond between family atoms. As an outsider, the US military can also form a small society. But this type of military society belongs to a typical "social-oriented society" structure. The US military therefore emerged as an external member of the entire South Vietnam family-oriented society. In this way, the interaction between the US military and basic society member of South Vietnam will also be smaller than the interaction within and outside the family atom of South Vietnam society. This is why the US military is completely isolated from the entire South Vietnamese society. The reality is that after 20 years of hard fighting, the U.S. finally had to completely withdraw from Vietnam.

Of course, this may actually be the case within a family-oriented society. For example, during the period of dynasties in China, a small social-oriented society would briefly form within the peasant rebel army. If these troops were merely fighting, the outcome would not have been much better than what the U.S. military did in the Vietnam War. But in China's successful peasant uprisings over the years, a family-friendly approach has been adopted. For example, methods such as "beating local tyrants and dividing lands" are used to meet the needs of ordinary families to obtain energy. At the same time, while fighting, they help ordinary people to do farm work, housework, etc., and even the army reclaims wasteland and grows food crops by itself (Lei, 2013). Such a series of measures acts like a household "ionic bond". This has greatly shortened the relationship between the army and the family atoms, and improved the social interaction between the army soldier and the family atoms.

It can be compared with the US military presence in Germany. Since the United States and Germany are both balanced societies, although the number of American troops stationed in Germany is as high as 70,000, it is easier for these American troops to integrate into German society. The United States announced in 2020 that it would reduce the number of US troops in Germany, and less than 50% of the German public expressed support. Let's take a look at the US troops stationed in Japan. There are more than 30,000 land combatants and tens of thousands of family members. Although Japan is also a family-oriented society, after the Meiji Restoration, Japan is stronger than other Asian countries in accepting Western culture. Therefore, the US military stationed in Japan will not cause a conflict as serious as the Vietnam War. However, US troops stationed in Japan often have serious conflicts with local people (Chanlett-Avery & Rinehart, 2012).

References

Bakunin, M. A., & Walter, N. (1971). *The Paris Commune and the idea of the state.* CIRA.

Buckland, M. K. (1991). Information as thing. *Journal of the American Society for Information Science,* 42(5), 351–360.

Chanlett-Avery, E., & Rinehart, I. E. (2012, August). *The US Military presence in Okinawa and the Futenma Base Controversy.* Library of Congress Washington DC Congressional Research Service.

Cheng, Z. (2019). *Cognitive development theory.* American Academic Press.

De Vries, A. (2018). Bitcoin's growing energy problem. *Joule,* 2(5), 801–805.

Delpit, L. (2006). *Other people's children: Cultural conflict in the classroom.* The New Press. 170–172

Gruber, R. E. (2002). *Virtually Jewish: Reinventing Jewish Culture in Europe*. University of California Press.

Hopfield, J. J. (1982). Neural networks and physical systems with emergent collective computational abilities. *Proceedings of the National Academy of Sciences, 79*(8), 2554–2558.

IEA (2019), Bitcoin energy use – Mined the gap. IEA, Paris. https://www.iea.org/commentaries/bitcoin-energy-use-mined-the-gap

Jupp, J. (1997). Creating multicultural societies: Australia, Britain, Sweden, and Canada. *International Journal*, 52(3), 508–523.

Lei, Z. (2013). Historical origin and realistic enlightenment of" Nanniwan Policy"-On the realistic and pragmatic scientific spirit of Zhu De. *Asian Culture and History*, 5(1), 13.

Lindong, Q. U. (2008). The development of Chinese ethnic minority historiography and the formation of the concept of a united multiethnic nation. *Frontiers of History in China*, 3(2), 165–172.

McGarry, A. (2014). Roma as a political identity: Exploring representations of Roma in Europe. *Ethnicities*, 14(6), 756–774.

Moyar, M. (2006). *Triumph Forsaken: The Vietnam War, 1954–1965*. Cambridge University Press.

Nickbakhsh, S., Mair, C., Matthews, L., Reeve, R., Johnson, P. C., Thorburn, F., ... & Murcia, P. R. (2019). Virus–virus interactions impact the population dynamics of influenza and the common cold. *Proceedings of the National Academy of Sciences*, 116(52), 27142–27150.

O'Connor, A., & Batalova, J. (2019). Korean immigrants in the United States. SPOTLIGHT. MPI Migration policy institute. https://www.migrationpolicy.org/article/ onfuc-immigrants-united-states-2017

Piaget, J. (1964). Part I: Cognitive development in children: Piaget development and learning. *Journal of Research in Science Teaching*, 2(3), 176–186.

Prigogine, I. (1978). Time, structure, and fluctuations. *Science*, 201(4358), 777–785.

Shannon, C. E. (1948). A mathematical theory of communication. *The Bell System Technical Journal*, 27(3), 379–423.

Traub, J. F., & Woźniakowski, H. (1990). *Information-based complexity: New questions for mathematicians*. International Computer Science Institute. 34–43

Von Mises, L. (1981). *Socialism*. Ludwig von Mises Institute.

Weiming, T. (2017). *Ecological implications of Confucian humanism*. Faith Values and Education for Sustainable Development, 78.

3

Interactions in the social systems

3.1 Interaction between individuals

3.1.1 Social interaction and gender interaction

From the analysis in Chapter 2, information transfer can cause interaction. Specific to human society, there are gender differences. The existence of this gender difference is also the key to the stable structure of human society. It is just like the physical world is composed of two microscopic particles with different electric properties, electron and proton.

What is conveyed in social interaction is the factual information about the existence of various things, which is used to describe the process of developing and ending something. Of course, because the genders exist objectively, some gender relations will also be involved when describing information about various objective facts. Therefore, the boundary between social information and gender information is not so clear.

Generally speaking, gender interaction is happened between two different genders, mainly through the interaction caused by the exchange of sex hormones. Gender information is passed in the process of gender interaction. Essentially different from social information, gender information involves sex hormone transmission.

In order to distinguish two different interactions, we can use different symbols to represent these two interactions. As shown in Figure 2-3(a). For social interaction, the social coupling coefficient caused by the transmission of information is represented by a. The coupling coefficient of gender interaction is represented by g.

The difference between social interaction and gender interaction can be compared with gravitational interaction and electromagnetic interaction in the physical world. Social interaction is similar to gravitational interaction. The gender interaction can correspond to electromagnetic interaction due to the existence of two different genders.

Although electromagnetic interaction is also a macroscopic interaction, its strength is far greater than that of gravitational interaction, and usually two different charges are tightly bound together on a microscopic scale. For example, electrons and protons are bound together to form an atom. Gender interaction usually occurs only when the two individuals are relatively close in time and space. Like gravity, social interaction is a kind of ultra-remote interaction, which can cover a very large range, and social interaction can be generated as long as the information can be transmitted. Once a society is formed, it can be either a small one, like a primitive tribe in the Amazon jungle, or a very huge one, like the former Soviet Union.

Of course we can also notice an important difference between social and gender interactions and gravitational and electromagnetic interactions. This difference is mainly manifested in the application of the law of conservation of physics. Social and gender interactions convey information and need to continuously absorb energy and momentum from the environment. The gravitational and electromagnetic interaction can be an isolated system, maintaining the conservation of energy and momentum. Of course, the mechanism for generating the interaction will also be very different.

Of course, the information interaction caused by information transmission in human society and gravitational interaction will be different.

On the one hand, due to the diversity and complexity of information transmission, information interaction will be more complicated. On the other hand, considering that the information interaction has only two dimensions, time and state, the method to deal with social problems will be simplified again.

3.1.2 The impact of information technology on society

If various modern communication technologies are considered, the speed of dissemination of information will become faster and wider. This is the same as the effect of electromagnetic waves propagating in different media. Information sent directly to the space at this time can be ignored. Information basically travels on a two-dimensional plane. So the introduction of information technology has led to great changes in the state of communication in the entire society.

The massive application of information technology has also led to an unprecedented development in the creation of information. The total amount of information has greatly increased. According to data, the energy consumed by the Internet is increasing exponentially (Hinton, Baliga, Feng, Ayre, & Tucker, 2011). And our previous analysis has shown that energy consumption is proportional to the amount of information.

This shows that the large-scale application of information technology, on the one hand, leads to the rapid growth of the total amount of information. On the other hand, the application of information technology should make the dissemination of information change from three-dimensional space to two-dimensional space in human society. This change leads to a flattening of the social living space, that is, from an approximate three-dimensional space into a completely two-dimensional information space. This will affect the quantitative calculation process of the information space.

3.1.3 Social interactions

Information interaction covers a wide range. Even the basis of gender interaction is the transmission of information, but the main transmission of gender interaction is hormones and other physiological information. When studying social systems, it is more convenient to divide interactions into social interactions and gender interactions. But the difference between them is not either/or. In fact, social interaction can also be more closely related to gender interaction.

If it is a social oriented society, the main form of social interaction is information interaction. In the whole society, the interaction between basic social members is mainly based on information exchange. This is why many social oriented societies attach great importance to the control of information dissemination.

But in other types of societies, in addition to including information interaction, other forms of interaction are more often used. For example, in family-oriented society, social interaction mainly adopts the way of gender interaction. In a family-oriented society, very strong social relationships such as "covalent bonds" can be formed using gender interactions. This also enables family-oriented society to form stronger social interactions than social oriented society in the condensed state. And the condensed matter makes the distance between the members of society pulled in. This enhances not only the role of covalent bonds, but also the information interaction within the family-oriented society. Therefore, under the conditions of small social space and high population density, the social interaction in family-oriented society is significantly stronger than that in social oriented society.

Although social interaction in the family-oriented society is closely related to gender interaction. But this covalent bond is still different from the gender interaction between men and women. In covalent bonds, gender interactions are encapsulated. Rather, it extends high-intensity gender interactions into social interactions for information exchange in a manner similar to a temperature gradient.

On the other hand, the information is more suitable for long-distance transmission. In particular, the application of information technology allows information to spread farther. Due to the limitation of physiological factors, the action distance of gender interaction is much shorter. Therefore, when the social space is large enough, the distance between basic social members will expand, which will significantly weaken the strength of gender interaction. It also weakens the information interaction between family members. When gender interaction weakens, it naturally weakens the strength of this "covalent bond" connection between families. As a result, social interactions in family-oriented society weaken rapidly when the social space increase. Under the same conditions, the social interaction among the basic social members in a social oriented society will not weaken too much. At this time, social oriented society will become more stable than family-oriented society.

We can see this from the population densities of Tokyo and London. As a family-oriented society, social individuals within it tend to gather together in smaller social spaces to enhance social interaction. This results in Tokyo's high population density. As a balanced or social oriented society, the social individuals within it are less sensitive to the size of the society. This results in cities like London having a much lower population density than Tokyo.

3.1.4 Gender interaction

Gender interaction is also generated by exchanging information. But unlike other types of interactions, gender interactions mainly transmit information through sex hormones. Sex hormones are mainly chemical substances synthesized by organs such as the gonads of living things. These chemicals can promote the development of sexual organs, and then transmit the message of gender interaction to each other.

Since the secretion of sex hormones is always associated with a specific biological gender, gender interaction usually occurs only between two different genders and forms a close connection.

To better compare gender interaction with electromagnetic interaction, we also assume the existence of a gender field here. That is, each individual gender can form a gender field similar to the electric field around him or her. Therefore, as long as the field exists, it can attract individuals of the opposite sex, and repulsion to same-sex individuals.

Considering the electromagnetic interaction, the movement of electrons can form an electric current. To create the movement of electrons, an electric field can be applied to the electrons. In the process of gender interaction, a phenomenon similar to the movement of electrons driven by an electric field can also be formed. Usually, when an individual is in the gender field of the opposite sex, there will be two individuals moving close to each other.

For animals, this law of attraction of opposites is similar to the mutual attraction of positive and negative charges. But human gender interaction patterns are more complex than animals. After all, in human society, there are more complex information interaction methods to limit the generation of this human gender interaction.

For animals and even humans, gender interactions are also affected by other factors. One of the important factors is the value of the individual gender. The value is mainly reflected in the extent to which the needs of the other party can be met. Specifically, it is reflected in the strength of the interaction between an individual and the gender field. If a gender individual has a greater effect on the gender field, the greater the gender value, the easier it is to attract other opposite sex. The gender value of animals is mainly reflected in the mating ability. The gender value of higher-ranked humans is more affected by factors such as physical environment, cultural background, and economic ability.

It can be seen that, unlike social interaction, the intensity of gender interaction has little to do with information technology. Gender interaction is mainly affected by physical conditions, economic conditions, and humanistic conditions. When the physical environment, human environment and other conditions are basically the same, the economic conditions will become the main conditions. That is to say, the economic conditions are good, which means that it will help to improve the strength of gender interaction. It is easier for gender flow to areas with better economic conditions.

3.1.5 Case study: Gender flow in East Asian countries

Due to the long history and the very frequent population movements in the four countries, the four countries of China, Japan, South Korea and Vietnam have many very similar factors. These factors include the same ethnicity, the same cultural background, and roughly the same physical environment. Therefore, the main difference between these four countries is still in the level of economic development.

From the perspective of the per capita gross domestic product (GDP) of the four countries, Japan and South Korea are roughly the same, and both belong to developed countries. The economic development of developing countries China and Vietnam is relatively lagging behind. In addition, Vietnam's per capita GDP is lower than China's. Therefore, according to the direction of gender flow, the gender flow of women mainly flows from Vietnam to China, Japan and South Korea, and China to Japan and South Korea.

Judging from the official figures released by various countries, it basically confirms the direction of such a flow. According to the survey results of the Korean Ministry of Justice in September 2018, as of 2018, there were 3,671 Chinese brides and 6,338 Vietnamese brides in foreign-related marriages in South Korea (Hollingsworth, Seo, & Bae, 2020). During the same period, the number of Chinese and Vietnamese grooms was much smaller. According to the statistics of the Ministry of Health and Welfare of Japan (Japan Data, 2018), among the cross-border marriages in Japan in 2016, Chinese grooms accounted for 12.5% (790 people), and brides accounted for 33.4% (5526 people). It can be seen that the flow of Chinese brides to Japan is still the main gender flow. For South Korea with similar economic development, the proportion of Korean brides who married to Japan in 2016 was only 13.7% (2031), and the proportion of Korean grooms reached 25.7% (1627).

Of cause, trends in gender flow can also be influenced by other factors.

Another comparison is that interracial marriages in Japan only account for 3.7%. The proportion is very low. It shows that other factors such as cultural background have a considerable influence on the number of cross-border marriages in Japan.

Neither China nor Vietnam have statistics on cross-border marriages between the two countries. However, in the "China Civil Affairs Statistical Yearbook", the overall data on cross-border marriages are counted (Ministry of Civil Affairs of the People's Republic of China, 2020). According to this data, the total number of cross-border marriages in China in 2019 was 48,424, of which the total number of foreign brides who married into China was 29,329. The number of Chinese brides married abroad is 19,095.

It can be seen from here. The number of foreign brides who marry into China is nearly 10,000 more than the number of brides who marry from China. If you remove the small amount of data on marriages with Korean and Japanese women, it can be determined that these brides who marry into China are mainly distributed in Southeast Asia. Due to the large population of Vietnam, it is believed that the number of Vietnamese brides marrying into China accounts for the majority of the total.

Then we compare the economic development of the four countries. In 2020, China's per capita GDP will reach US$10,400. Japan's per capita GDP reached 40,200 US dollars. South Korea's GDP is $31,600. It can be seen that the economic development level of Japan and South Korea is similar, and Japan is slightly higher. However, the economic development level of Japan and South Korea far exceeds that of China. In 2020, Vietnam's per capita GDP was only US$2,785, which is far lower than the other three countries. Combining the direction of the flow of brides in the four countries, it can be found that it is basically consistent with the economic development of the four countries.

In addition, it can be noticed that there is also a trend of female migration to developed areas and rural to urban migration within China. The cities where these gender flows flow mainly include Beijing, Shanghai, Guangzhou and Shenzhen.

This all reflects that the factors affecting human gender interaction are complex. Not just limited to physiological factors.

3.1.6 Statistical mean of interaction

Although we constructed the corresponding social interaction and gender interaction models by comparing gravitational interaction and electromagnetic interaction. this interaction due to the information field is very different from the physical interaction. This is the social and gender interaction coupling coefficient individual differences are very large. Especially in social information interaction, the amount of information produced by each basic member of society is closely related to his growth environment, education level and information technology mastered. This makes it almost impossible to find individuals who can produce two identical amounts of information in the entire society. Even in the gender coupling coefficient, there are very large differences. By comparison, the gravitational and electromagnetic interaction both have very fixed gravitational and dielectric constants.

In order to be able to deal with social problems more effectively. We can construct a more ideal social model. In this model, we assume that the interactions between all social base members, including the coupling coefficients of social interaction and gender interaction, are exactly equal. This makes the problem much simpler. And more importantly, we can directly apply various laws and research results in physics to the process of social development and change.

We can take an average value of various interaction coupling coefficients in the actual society as the social and gender coupling coefficient in this "ideal social model". Calculations for statistical averaging can use a number of different data distributions. Currently the most commonly used is the normal distribution.

Math further

The probability density of a normal distribution is usually of the form

$$p(\gamma)=\frac{1}{\sqrt{2\pi}\sigma}exp\left[-\frac{(\gamma-\overline{\gamma})^2}{2\sigma^2}\right] \tag{3-1}$$

Where, $\overline{\gamma}$ is the expected value or mean of the distribution, and σ is the standard deviation of the distribution.

If this normal distribution is applied to social interaction, we can call $\overline{\gamma}$ the social expectation or social average. The standard deviation σ is called the social standard deviation.

Different types of societies have different averages. In addition, for a society, if all basic social members have the same cultural background, the social standard deviation σ will be relatively small. Therefore, we can also see some important characteristics of different types of societies through the formula of the normal distribution.

For family-oriented society, the average value of the social interaction coupling coefficient related to information transfer is relatively small. The more single the ethnic composition of the whole society is, the smaller the social standard deviation σ will be. In this way, we can infer that the social standard deviation of China will be larger than that of Japan. This is because the ethnic composition of Japanese society is relatively simple.

For a social-oriented society, the average value of the social coupling coefficient related to information transfer will be much larger than that of a family-oriented society. Like family-oriented society, if the ethnic composition of the whole society is more homogeneous, its standard deviation will be smaller. For example, the ethnic composition of the former Soviet Union is very complex, so its social standard deviation is very large. And now the social standard deviation in Belarus will be much smaller. However, in social-oriented society, the average social coupling coefficient related to gender interaction is smaller than that in family-oriented society.

For a balanced society, the average social interaction coupling coefficient is between family-oriented society and social-oriented society. Of course, due to the different cultural backgrounds of different countries, their social standard deviations are also very different. For example, the United States, as a diversified and balanced society, has a significantly larger social standard deviation than the United Kingdom. However, due to the complex ethnic composition of some countries in the European Union, the social standard deviation of most countries may be larger than that of the United Kingdom.

For gender interaction, it should also obey the normal distribution. But gender interaction is mainly related to the biological structure of the organism. The differences between human individuals are very small, so the mean and standard deviation of gender interactions for different types of societies should be the same as possible.

3.2 Gender nucleus in family atom

3.2.1 Gender relations

Electromagnetic interaction exists between electrons and protons, so that electrons and protons can combine to form a stable atomic structure. There is also a gender interaction similar to electromagnetic interaction between two different genders of biology. Therefore, two different genders can be combined together to form a stable structure. We can call this structure of two genders a "gender nucleus", because this nucleus may become the core part of the family atom.

In order to maintain the stability of the entire nucleus, we can assume that this gender interaction also has two kinds of energy, kinetic energy T and potential energy V. In this way, when two different gender individuals reach the equilibrium of kinetic energy and potential energy at the same time, the entire gender atom will be in a stable state.

Of course, to achieve this equilibrium state, it is necessary to satisfy a most basic "energy conservation", that is

$$E = T + V \tag{3-2}$$

As pointed out earlier, gender interaction is essentially an interaction caused by information exchange. Therefore, this "energy conservation" is actually a kind of "information conservation".

That is to say, the information potential energy of two different gender individuals is very high before the gender nucleus is formed, and the corresponding information kinetic energy is relatively small. After the nucleus structure is formed, the information potential energy decreases, and the corresponding information kinetic energy begins to increase. After the two reach equilibrium, the entire gender nucleus is stabilized.

Different from electromagnetic interaction, gender interaction is mainly affected by the individual's physiological characteristics. So although there can be information conservation in short time, with the process of individual physiological growth and decline, this information kinetic energy T and information potential energy V will change continuously with the growth of age. The strength of gender interactions will also vary widely for each age group.

We now know that the nucleus in the physical world is made up of neutrons and protons, and neutrons and protons have about the same mass. But neutrons are not charged, and neutrons can release a proton and an electron through beta decay. So neutrons can be thought of as the product of protons and electrons bound together in some way.

Of course, the mass difference between electrons and protons is very large, or the status is different. Unlike the nucleus in the physical world, in the gender nucleus, although gender interaction is equal, it may lead to unequal status of the sexes in the process of socialization. The emergence of this unequal status will form a situation like the difference in mass of electrons and protons. From the analysis of quantum physics theory, this proton has a very large mass, while the electron has a very small mass, which can allow the nucleus or atom to obtain a more stable structure.

From the perspective of the development history of human society, the earliest primitive society belongs to the matriarchal society. In such a society, women become the core of the entire gender nucleus. But after the Iron Age, males gained economic dominance due to the use of tools, which led to a shift in the core of the gender nucleus in society from females to males.

From this analysis, it can be seen that the economic power in human society has become one of the sources of the gender difference in "mass" in the gender nucleus, which is similar to the mass difference between protons and electrons.

But when we observe the society composed of animals, we will find that there are some differences from human society. For example, in a wolf pack, gender is on an equal footing. In a wolf pack society, a pair of male wolves and female wolves are the dominant wolves. But in African lions, a male lion is the core of the entire African lion and plays a dominant role. This shows that in the composition of gender atom, in addition to the more complex economic conditions in human society, there are other natural environments and conditions that play an important role.

By comparing the differences between wolves and lions, we can find that the differences between the two are mainly reflected in a so-called "social" aspect. That is to say, wolves are social animals, and they are better at gathering together for various information exchanges with their fellows. The social nature of lions is much weaker. Although lions live in groups, the relationship between lions and their peers tends to be more of a competitive relationship. This can branch from wolves that were easily domesticated by humans and gradually evolved into dogs. The domestication of lions is difficult.

The social nature of this animal leads to the fact that even if the sexes are in an equal state, more animals without gender interaction can be linked together because of the existence of the social coupling coefficient, forming an animal group society. However, animals with relatively weak sociality, like African lions, because the social coupling coefficient is very weak, it is impossible to maintain the relationship between animals without gender interaction and form a relatively solid social structure. It leads to the gender competition in the Africa lion pride, and some related male lions become the nucleus of a lion pride in the end.

It can also be seen from this that males become the core of the gender nucleus due to the difference in economic strength of human beings, which is essentially caused by a competitive relationship between males. This is basically the same as the formation of African lions.

Here, the concept of Higgs field in physics can be used to deal with the situation where the gender individual has a competitive advantage. According to the derivation of the Higgs field, we can also regard this gender field as a kind of field that can give "value" to the individual gender, then as long as the gender interaction exists, these individuals are always interacting with this gender value field. If a gender individual interacts more strongly with the gender value field, its value is higher, because it can better meet the needs of the other gender. The higher the value of the gender individual, the more attractive it is, just like the greater the mass of the material, and the more energy is required to move it. If the interaction between a gender individual and the value field is relatively weak, then its value will be lower. This is equivalent to a lower mass object requiring less energy to move it.

So in this case, we can find some more interesting phenomena. If in a social or animal population, the males must compete to obtain gender interaction with the opposite sex. This will lead to a larger difference in the gender value of male individuals.

This means that in the male group, there are a small number of high-value males who can gain the power to engage in gender interactions with females. However, a large number of males with relatively low value appear to be unable to obtain gender interaction. But this is not the case for females. For females, because of the lack of this competition, their interaction with the gender value field is basically the same for all females. So the difference in value across all females will be smaller. As a result, when the gender atom is combined, a male may interact with multiple females at the same time to form a gender nucleus.

If it is reversed, such as in some social systems composed of certain insects, only females are fertile, and only stronger females can produce more children, this can lead to competition between females. Some of these females are able to interact more strongly with the gender value field and thus obtain higher values. Relatively speaking, in the absence of male competition, the difference in gender value of males is relatively small, and they are basically at an average level. In this case, there may be a gender interaction between one female and multiple males, and finally constitute multiple body gender nucleus.

Therefore, in social groups composed of different organisms, gender interaction still has a certain degree of complexity. Due to the influence of the interaction between the individual gender and the value field, different genders have different values. The difference in the individual value of the gender will lead to a very large difference in the structure of the gender nucleus.

3.2.2 The formation of basic social members

In a social-oriented society, the nucleus formed by the relationship between the genders can directly constitute the most basic unit of the whole society. Therefore, this gender nucleus can be regarded as an independent social molecular in the social-oriented society.

If it is in a family-oriented society, on the basis of the gender nucleus, once the offspring are reproduced, these offspring children will surround the parent's gender nucleus to form a structure like the atom in the physical world. This family atom can be seen as a social atom. In family-oriented society, most of the basic society members consist of this family atom and very few of them are gender molecular.

Therefore, the basic social members of a social-oriented society are the simplest. Since the social molecules is also the gender molecules, the gender molecules and social individuals make up all the basic social members.

In a balanced society, the proportions of family atom and gender molecular are similar. For a balanced society, the gender relationship formed by the gender interaction is not yet the most basic member of the entire society sometimes. In balanced societies, the gender nucleus can also form the family atom through complex intra-family social interactions.

In terms of the internal structure of the family atom, in the family-oriented society, in the gender nucleus, there is a one-to-one gender of males and females, as well as a one-to-many or many-to-one gender. One-to-many situations can be further divided into male-dominated or female-dominated situations.

So the relationship between gender nucleus, social molecular and family atom is similar to gas molecular and liquid, solid atom or ion etc. in the physical world. Gender molecular is similar to gas molecular. Such as hydrogen, oxygen molecular and so on. The family atom is similar to an atom with a relatively large atomic weight. Such as iron, copper, gold and other atoms or ions.

In addition, it can also be noted that there is also a problem of individual growth and development in a society composed of humans or other animals. Gender interaction of humans or animals only occurs in a specific age period. During this period, the physiological development of human or animal gender-related organs is the most mature, and the gender interaction will be the strongest. Especially when humans are in childhood, elderly and even middle-aged, gender interaction is very weak. In this case, maintaining the stability of the social molecular must rely on other constraints to achieve the goal. These constraints mainly involve the cultural, ethical, moral and other factors of a society. Cultural factors are derived from social interaction. It can be seen that, especially for human society, even gender interaction is affected by social interaction such as cultural ethics and morality.

That is to say, family-oriented society and balanced society are formed under the condition of mutual interaction between a family and society. This is different from the atom or molecular in the laws of physics, which can be formed entirely by electromagnetic interaction.

3.2.3 Case study: Dynamics of African lions

Among the cats, the African lion is the only gregarious grouping animal. The reason why I am interested in African lions is because their grouping is mainly related to gender interaction.

It is generally believed that African lionesses live in groups for better hunting purposes. However, scientists have found through research that, in fact, for strong animals such as lionesses, the hunting effect of a single lioness is much better than that of several lionesses at the same time. And after a successful hunt, a lioness can eat more food.

Therefore, after research, it was found that the lioness's group behavior is mainly to protect the cubs (Packer & Pusey, 1997). These cubs are born both by themselves and by other companions in a ride. A lioness is only in heat for a short period of time. In this way, the number of lionesses with gender interaction ability in a lion pride is not many, and the average is relatively small.

In contrast, although the estrus period of male lions is not long, the estrus period of each male lion in two or three years is relatively far longer than that of female lions. In this way, male lions can conduct gender interaction longer than female lions. There will be an imbalance of gender interaction. And lionesses like to live in groups to form a pride, so that they can help each other in raising their offspring. So on the surface it seems that in a ride with more than a dozen female lions, only a few male lions participate in gender interaction.

From the point of view of the value field, the small number of male lions in a lion pride is of course also related to the gender value of male lions. We can note that the need for gender interaction is greatly reduced as long as the lioness has produced offspring during the lifetime of the African lion. In addition, the lionesses gather together to form a pride, not because of the needs of gender interaction, but mainly because of the need to raise offspring and provide a safe shelter for themselves and the offspring of other lionesses in the pride. Therefore, it can be seen that the interaction strength between the lioness and the gender value field is generally relatively weak.

But male lions spend their entire lives looking for female lions to mate with, and even these males rarely hunt on their own. This also means that the male lion has a stronger interaction with the gender value field. This interaction causes some of the male lions to have a higher gender value, while other male lions have a weaker gender value in the competition.

This kind of interaction between male lions and gender value field is usually manifested in several male lion alliances that are only related, and have strong mating ability. On the other hand, in the process of competing with other unrelated male lion alliances, it shows stronger physical fitness and cooperation ability. The last is the hostility and expulsion of the lionesses faced when invading a pride. Of course this can also be called a process of natural selection. From the analysis of the field theory in the physical theory, we can also regard the various challenges faced by this lion in his life as a gender value field. In the process of the male lion interacting with such a gender value field, the gender value of a male lion can be fully displayed.

Even if a lion pride has been successfully invaded, individuals in the lion union actually have gender values. It can be found from the research that although there are several male lions in the lion group, DNA tests have found that most of the offspring are derived from only one male lion. This shows that gender values vary widely among male lions.

As for the gender value of the lioness, the gender value is relatively average, and the difference is not obvious. After all, the size of a pride is mainly determined by the distribution of resources in the region where the lions are located (Mbizah, Valeix, Macdonald, & Loveridge, 2019). If the gender values of the lionesses vary widely, the size of these lion prides will be related to the competing relationship between the lionesses. Like male lions, only competition can create problems of individual or group differences.

From the point of view of natural selection, lions with high gender value are still very beneficial to lion pride. Including can better prevent the invasion of other males, so as to protect the pride to better raise offspring. In addition, male lions with higher gender values can also gain a genetic advantage.

3.3 Family atoms

3.3.1 Kinship

The formation of gender molecular also means the formation of an independent basic society members. As part of a society, it can exchange information with other basic society members.

On the other hand, biological individuals within the gender nucleus have reproductive capacity, and can have offspring through gender interaction. Although there is no gender interaction between viruses and bacteria, for viruses, mRNA can guide DNA synthesis, thereby replicating individuals that are completely identical to themselves. For bacteria, more individuals can be divided by cell division. Therefore, the formation of a society is always accompanied by the process of continuous growth of individuals in it. The emergence of new individuals means the emergence of new social interaction and gender interaction.

Here there is a problem of interaction between individuals with kinship. The gender nucleus as a whole can naturally communicate with its own offspring. This information exchange is consistent with social interaction. But we can also notice the difference between them. The intensity of information exchanges between parents and their children is significantly greater than that with other members of society. This is because parents and children are closer in time and space. Most of the knowledge children receive about the world is passed on from their parents. This also means that the cognitive structure of parents and children in the neural network system of the brain is basically the same. This consistency shows that the cognitive coupling coefficient between the two is obviously stronger than that of other social members.

This phenomenon of stronger cognitive coupling coefficient between members with kinship is widespread even in the animal kingdom. For example, in the lion alliance in the lion pride, brothers with kinship are more likely to form such an alliance. Cooperate more fully with each other, and it is easier to gain an advantage in the process of competing with other Lions leagues. Even after having occupied a lion pride, these male lions do not care that the "brother" in the alliance mates with the lioness (Packer & Pusey, 1997).

But a stronger coupling coefficient does not mean that parents and children will naturally combine into a family atom that is as strong as the atomic structure of matter. This is because after the children grow up, if they receive more information from external social members, the children will live independently from their parents.

This biosocial group effect is evident in the dynamics of wolves. The current dynamics of the wolf pack suggest that the entire wolf pack is a biological social group with parents as the core (Mech, 1999). This group is relatively stable and solid. But after the offspring of female wolves are adults, these adult wolves will go out alone to find other heterosexual mates and form a new wolf pack.

Therefore, in order to maintain the integrity of the whole family even after the children become adults, other biological conditions and even human-specific ethical and moral constraints are required.

3.3.2 Formation of family atom

Due to kinship, the social interaction coupling coefficient among family members is significantly larger than the coupling coefficient with other social members. At this time, family members can form stronger bonds. If the relationship between family members and other family members in the society always maintains a relatively weak connection, or if the ethical and moral requirements of the whole society strengthen the connection between family members, then the whole family can become a basic member of society. This is the basis for the formation of family-oriented society.

Considering that the basis for the formation of family members is the gender nucleus composed of different genders. Then the gender nucleus expands the size of the family by multiplying offspring, thereby forming the structure of the family atom. Therefore, the structure of the family atom will also have various forms. There are two main ones, the gender-equal family atom and the gender-unequal family atom. Of course, they can also be divided into monogamous families, polygamous families or polyandry families according to the family forms that exist in human society.

The "family" composed of two different gender ratios can be represented by the Feynman diagram shown in Figure 3-1.

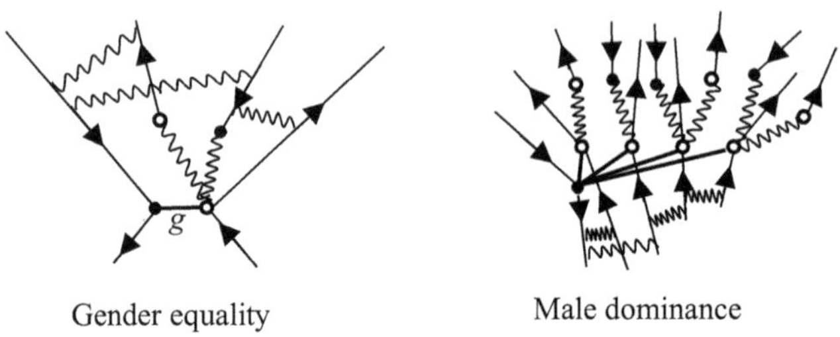

Figure 3.1. Two different sex ratios in the family structure

As can be seen from the left side of Figure 3-1, if the two genders are equal to each other, such interaction and reproduction of offspring, the interaction of the entire family or society is relatively simple. In the figure, two offspring can be produced through gender interaction between the father and the mother. After the two offspring are reproduced, both the father and the mother can exchange information with the offspring through social interaction at basically the same status. If this information exchange is reinforced by some kind of moral and ethical constraints of human society, the social coupling coefficient among the four members can become very large, thus maintaining the integrity of the whole family or society.

For the male-dominated situation on the right side of Figure 3-1, a male and multiple females are coupled together through gender interaction, forming a relatively strong connection. After the female reproduces offspring, the connection between the children and the mother will be relatively close, and there will be relatively close communication between them. But fathers only have close social interaction with their mothers, and social interactions with their children will be weaker.

From the perspective of gender interaction, such polygamous families are less stable than monogamous families. The main reason is that, on the one hand, like the African lion pride, men need to compete to get enough female spouses. This means that the gender value difference between men can become very large. This will lead to very fierce competition between men. Consider the complexity of human cognitive abilities. The uncertainty of this competitive outcome will become very large. On the other hand due to the large population, even with the moral and ethical constraints of human society, such a family or society will not be stable. It is more likely that it will eventually split off and become multiple families.

As can be seen from Figure 3-1, although separated from social constraints, family atoms can form monogamy, polygamy, polyandry and even other more complex structures, but under the constraints of social ethics, law and economic ability, the family atom of human beings is mainly monogamous at present.

In the animal kingdom, wolves tend to have a gender-equal family atom structure. In a wolf pack, there is usually an alpha male and an alpha female as the parents of the whole pack, and after producing enough offspring, a wolf pack family is formed. The Africa lions tend to have an unequal gender family atom structure. In a lion family, there are usually one to several male lions, together with a dozen female lions to form a family. This is similar to the polygamous family form of human society. However, compared with the wolf family, the family structure of the lion is not stable, and it often faces challenges and invasions from other male lions outside.

Compared with the animal family atom, the structure of the human family atom is much more complicated. At the same time, there will be more factors affecting the individual interaction within the human family atom. In addition to gender interaction linking parents together, kinship is also an important factor in maintaining the structural integrity of the entire family atom. This is common in both the human and animal kingdoms.

However, there are more complex information exchange and preservation technologies in human society. This complex exchange of information can convey more complex information. And knowledge preservation methods such as written books can pass on human culture. Therefore, after the accumulation of human cultural knowledge to a certain extent, the constraints of ethics and morality unique to human society are formed. This kind of human ethical and moral constraints can strengthen the connection between family members, and can also weaken the connection between family members. This leads to great differences in the stability of the family atom structure formed in different human social forms.

Among them, like traditional East Asian countries, what is passed down is the requirement of moral restraint between members of the family. For example, in the ancient Chinese classic ethical works: The Great Learning, The Doctrine of the Mean, The Analects, and Mencius, the core is that everyone in society must respect their parents and love their children. Only when you handle your family relationships very well can you be qualified to take on the important responsibility of governing the entire society. In order to allow young people to receive the inheritance of this culture at a very young age, in terms of children's education, a considerable number of enlightenment books were published in ancient China. One of the most famous is "Three Character Classic ". In the book, it told the children the story of Huang Xiang, a little girl who knew how to use her body to help her parents warm the bed when she was nine years old. Through Kong Rong's 4-year-old, he knew to give pears to his elder brother to teach children how to be humble among brothers. As for the parents, it told the story of a father named Dou Yanshan who taught his children well, and all five of them eventually became talents. It can be seen that in ancient Chinese ethics, it is emphasized that family members with kinship should respect and love each other, parents should shoulder the responsibility of educating their children, and brothers and sisters should help each other. This kind of ethics is instilled into the neural network of children's brains from a young age, and finally forms the unique family culture of East Asia. This makes the family atom structure in East Asian cultures stronger and more stable. Even if the children grow up and separate out to form their own family atoms, these family atoms with kinship are maintained by the parents at the top, and they are essentially just a larger family atom.

Western culture is very different from traditional Chinese culture. In the children's enlightenment book "Bible Story", it tells the story of Adam and Eve being tempted by a poisonous snake and stealing the forbidden fruit. Explain that Western culture considers sex to be an original sin, and gender interaction is the cause of the problem. So this is a measure to weaken gender interaction. On the contrary, in traditional Chinese culture, it is an attitude of encourage gender interactions, after all, this is the basis for inheriting the lineage. In the story of

Moses leading the Israelites out of Egypt, the message to the children is that one person's primary responsibility is to contribute to the nation as a whole. But in traditional Chinese culture, it is believed that a person must first deal with his own family problems before he is qualified to talk about how to govern the country.

It can be seen from this that although Western culture also emphasizes the role of the family, it tends to weaken the gender interaction that forms the basis of the family, and emphasizes the importance of social interaction. This also leads to the obvious differences in the social forms of the East and the West.

3.3.3 Interaction between family atoms

Just like the molecular of the physical world. Once the family atom is formed, there is an interaction between the atoms. This interaction between atoms results in more families being brought together as a whole.

The interaction between gender molecules is mainly in the form of information exchange between gender molecules. The social interaction is formed through the mutual information exchange between the molecules.

For the family atom, due to the more complex structure, in addition to relatively weak information exchange, the direct and non-direct descendants outside the gender nucleus can be regarded as the periphery "free electrons" that can flow to other families outside the family at any time. If this family member forms a strong amphipathic bond with the corresponding gender in other families, it can be regarded as a covalent bond in an insulator. And if such free electrons can flow freely in the whole society, they can be regarded as free electrons in metals that can conduct electricity. Therefore, the social interaction in the family-oriented society is far more complicated than the social interaction in the social-oriented society. Social interaction in social-oriented society can be handled in terms of interaction between gas molecules. This can be solved directly using the Van der Waals equations. In family-oriented society, the strength of the social interaction coupling coefficient should also take into account the strength of interactions such as covalent bond and ionic bond in the physical world.

If we only look at the interaction caused by information exchange, on average, the total amount of information exchanged with the outside world in each person's life is roughly the same. We can infer that if in a society, the information exchange within the family is more closely, the amount of the information is larger, the information exchange outside the family will become weaker, and less information will be transmitted. The resulting is that the information interaction outside the family will be weaker. On the contrary, if the communication of information within the family is relatively weak, the communication between family members and the outside of the family will be closer. This also means that social interactions related to information exchange outside the family will be stronger.

However, if the influence of covalent bond and ionic bond of gender interaction is considered, since the number of social individuals of family atom is large, it is easier to form covalent bond and ionic bond with other family atoms. However, the gender molecular structure is relatively simple, and the stability is not as good as that of the family atom of more social individuals. Therefore, the conditions for forming a covalent bond or ionic bond between gender molecules are much more stringent. This is why most gases with relatively small atomic weights have low melting points. Like hydrogen, the melting point is only -259 degrees Celsius. Most substances with relatively large atomic weights have relatively high melting points. For example, the melting point of iron reaches 1538 degrees Celsius.

Therefore, in addition to the internal interaction intensity of family-oriented society is higher than that of social-oriented society, the external interaction with other families is also higher than that of social-oriented society. This forms an important distinction between family-oriented society and social-oriented society.

This social interaction is more like the interaction between organic molecules, so the Van der Waals constant between them is naturally higher than the Van der Waals constant of water, which is adopt to about 5–10 times higher in this book.

3.4 Interaction between societies

3.4.1 Nation and society

No matter what type of society it is, as long as the internal basic society members have undergone sufficient interaction, closer connections between members can be formed internally. This tight connection helps maintain the integrity of society.

This tight connection also means that the cognitive coupling coefficient among basic society members will become stronger over time. The continuous enhancement of cognitive coupling coefficient is directly related to the continuous interaction among basic society members.

Considering that after a society is formed, there are differences in the interaction among basic society members. Therefore in Section 3.1.6 we use the normal distribution for statistical averaging. A normal distribution has two important parameters, the mean and variance of the normal distribution.

If a society has undergone sufficient social interaction to form a more solid whole, the average cognitive coupling coefficient will be significantly improved. At the same time, the variance of the cognitive coupling coefficient in this society will also be significantly reduced.

We can therefore call a society a nation with a high internal average cognitive coupling coefficient and a small variance in the cognitive coupling coefficient. The mean and variance of a nation's cognitive coupling coefficient will be significantly different from other nations.

3.4.2 Relatively independent social thermodynamic system

After a nation is formed, it may form a relatively closed system. Due to the problem of differences between individuals, it is necessary to use statistical methods to calculate the system. The various thermodynamic parameters are the result of this statistical averaging.

After the nation is formed, it will also become a relatively solid whole. And such a whole will have thermodynamic properties that are different from other nations or societies.

The parameters that represent a thermodynamic system mainly include pressure p, volume V, and temperature T. In addition, parameters such as entropy S, internal energy U, Helmholtz free energy F, Gibbs free energy G and enthalpy H of the thermodynamic system can also be obtained from these parameters.

Like physical systems, social temperature, space, and pressure are the three most basic parameters for a society to form. These three parameters can be described by the thermodynamic ideal gas equation. As for some other parameters in the thermodynamic system, in general, entropy reflects the degree of chaos in the system, and of course it can also reflect the total amount of information in the entire social system. If the total amount of information in this society is relatively large, it means that the entropy of this society is very large. Of course, the amount of information is too large, and sometimes people feel that the whole society is more chaotic. But we should know that the size of the amount of information actually reflects the civilization of the society.

The internal energy of a society reflects the energy the society possesses. Social internal energy is related to the activity level of individuals in the whole society, and is also closely related to the energy supply of this society. For a society with a highly developed level of technological productivity, due to sufficient food supply, the energy input of this society will also be very large. Naturally, the internal energy of this society has changed a lot.

In addition to the above commonly used social thermodynamic parameters, we can also use parameters such as enthalpy and free energy to solve some problems of social thermodynamic state changes.

3.4.3 Case study: Disintegration of the Soviet Union

The Soviet Union was a very large social-oriented society. The Soviet society is composed of many ethnic groups and countries, and the compositional structure of basic society members is very complex. At the same time, in the Soviet society, the religious beliefs and cultural backgrounds of different ethnic groups are also very different. However, by weakening the interaction of members within the family, the Soviet Union managed to combine these peoples and countries of different cultural backgrounds and religious beliefs into a country with a large population and a vast territory.

From the point of view of Social Thermodynamics, the Soviet Union united these different peoples and countries into a huge social system, which involved the intermixing of multiple thermodynamic systems. When these thermodynamic systems are mixed with each other. The thermodynamic parameters such as temperature, entropy, internal energy, enthalpy and Gibbs free energy of the system all have very large changes.

The first is the change of social space. In the original nation-states, their territories were relatively small. After the integration of the Soviet Union, a social system with a very large social space was formed.

The second is the change in internal pressure. As the cultural differences between different ethnic groups are very large, and the thoughts of each basic society member are also different, this makes the exchange of information more frequent and individuals become more active. As a result, the internal pressure of the social system has become very large. And this increase in the internal pressure of the system led the Soviet society to seek outward expansion. This expansion includes not only the further expansion of national territory, but also the expansion of culture. This pressure to expand the territory and culture will eventually lead to the reaction force generated by other neighboring countries, which created additional external pressure on the entire Soviet society.

Of course, the integration of all national cultures in Soviet society also facilitated exchanges between different nationalities and cultures. Strong social interaction promotes the rapid development of more new ideas and science and technology. This cultural, technological, artistic and other development further promoted the rapid rise of the social temperature in the entire Soviet Union.

It can be seen that the three important parameters of thermodynamics, P, V, and T, have been greatly improved. This can be explained by the equation of state of an ideal gas.

But for such a huge society, its social space cannot be infinitely increased. After all, after the territory of the Soviet Union has expanded to a certain extent, it is impossible to expand it again. In addition, in terms of culture, although the Soviet Union has a diverse culture, in order to meet the requirements of the social-oriented society, the Soviet Union adopted the suppression of the development of national culture. At this time, the modern civilization that can replace these national cultures is still relatively insufficiency. For example, in terms of ideology, the Soviet Union adopted a set of rigid ideas and did not know how to test the truth through practice. In terms of culture and art, in order to weaken the interaction within the family, many taboo areas are set up, which stifles the creativity of literary artists. In the end, they can only seek development in technology. In the early days of the establishment of the Soviet Union, the development of science and technology did have a relatively good momentum. There have been a large number of Nobel Prize winners in science and technology. But in the late Soviet Union, rigid ideas and lifeless literature and art, coupled with a poor economy, eventually dragged down the development of science and technology in Soviet. In the late Soviet Union, at least, in terms of information technology, it can be said that the Soviet Union basically did not make very meaningful contributions to human beings.

Finally, according to the ideal gas equation, as the temperature rises, if the social space cannot be increased, its social pressure will rise rapidly. The increase in the internal pressure of the Soviet Union caused the reaction force of the neighboring countries to increase. Therefore, the pressure exerted by the neighboring countries on the Soviet Union also increased. This will lead to corresponding adjustment measures within the Soviet Union to slow down the rise of internal pressure. This mode of slowing is borne out by changes in the Soviet leadership over several generations. During Stalin's reign, through powerful measures, the interaction within the family was suppressed, and the entire society successfully turned to a social-oriented society. But the price of doing so is the gradual rigidity of thinking and the inability to develop culture and art. However, relying on the historical accumulation of various nationalities, the Soviet Union's technology has developed rapidly. After Stalin's death, Khrushchev took over the supreme power of the entire Soviet Union. In order to solve the various problems that had arisen before, Khrushchev made very big reforms in the political, economic and institutional aspects. These reform measures are aimed at promoting economic development and at the same time coordinating the contradictions between different countries and different ethnic groups. But these reform measures of Khrushchev touched the interests of some vested interests, which eventually led to Khrushchev's resignation. The entire Soviet society has returned to the original state of a relatively rigid system. In this relatively rigid state of the system, in order to prevent the continuous increase of social entropy, that is, the continuous increase of the amount of information in the entire society, measures such as cracking down on freedom of speech are used to restrict the social interaction of citizens and suppress the total amount of information in the entire society. Such an approach seems to be able to avoid social chaos and thus maintain the stability of the vast Soviet society.

But whether the social interaction is strong enough is the most effective means to maintain the society into a condensed state and become a whole, and to eliminate the ideological differences among the basic society members. The measures adopted in the late Soviet period to suppress freedom of speech resulted in the stagnation of the entire national economy and technological development, the inability to communicate effectively in society, and the weakening of social interaction, which meant that the entire late Soviet society was more akin to an ideal gas state. Compared to condensed states like liquids and solids, gaseous states are less stable. When subjected to external force, it is easy to be scattered. In the face of various internal and external pressures, the Soviet Union finally disintegrated in 1991.

After quantifying the thermodynamic parameters in subsequent chapters, we can directly use the Van der Waals equations to calculate the critical parameters of Soviet society in Section 6.3.7. In this way, the reasons for the collapse of the Soviet Union can also be supported by quantitative calculations.

3.4.4 Interaction between social thermodynamic systems

If there are multiple societies, there will be interactions between multiple social thermodynamic systems. Even if an isolated social system does not have any information, energy, temperature, etc. exchanges with other social systems, this social thermodynamic system cannot simply be regarded as an adiabatic system. After all, human society or other biological societies are always built in a complete ecological chain on the earth. All kinds of energy and information can be obtained from the earth's ecosystem at any time, and all kinds of energy and information can be released to this ecosystem at the same time.

However, we can also ignore other more complex factors, such as the transfer of energy and temperature of matter, and the influence of ambient temperature on the neural network of the brain, etc. If we only consider the interaction of information, then an isolated social system can also be approximately regarded as an adiabatic system. If we can think of a social system as an adiabatic system, we can use many of the tools of physical thermodynamics to deal with social problems. The solution to the problem can be greatly simplified.

If there are multiple social systems that are adjacent, and there is a strong exchange of information between these social systems, then any one of the social systems will be an open system at this time. There will also be interactions between these social systems, such as diplomatic activities between societies. Although this interaction is weak, it can link all these societies together into a whole. This is just as the ethnic composition in Southeast Asia is very complex, but all these ethnic groups can always coexist peacefully, because there has always been such a social interaction between them in history.

If multiple social systems are basically equal in size and strength, then through the interaction between social systems, they can be combined to form a larger social system. This larger social system can then be treated using the same thermodynamic treatment as a single social system. A typical example here is India. Although India is a very large country with a large population, it is actually composed of many independent national societies. There is extensive and long-standing social interaction between these different national societies, which is the most important reason why India can become a federation.

There is another situation, that is, if one of several social systems is a very large society, which has very strong advantages in terms of economic strength, technology and culture. In this case, to those small societies, the larger society seems to be a source of constant energy and information. These small societies can actually maintain a state of complete openness. In this completely open state, the energy and information of the whole society can be supplied indefinitely. This can maintain a basic balance in society as a whole. Of course, if there is a problem with this super-large society and can no longer provide these small societies with the resources they need, then these small societies may also be greatly dragged down by this, causing problems in the stability of the entire society.

This is the same as China's relations with neighboring countries. During China's thousands of years of historical development, these neighboring countries have always maintained an attachment relationship with China. It is also a social interaction that maintains the social stability and prosperity of these countries. But after the 18th and 19th centuries, the influence of Western culture caused strong turmoil in China itself. At this time, China cannot provide the necessary resources to these neighboring countries. Without an infinite supply of resources, these small societies themselves face great uncertainty. Among them, Japan has adopted a policy of comprehensive Westernization, vigorously introducing Western culture in an all-round way, which has enabled the entire Japanese society to maintain a stable state for a relatively long period of time. But some other countries around China, such as Vietnam, became French colonies. Later, for almost the entire 20th century, Vietnam faced the threat of various wars and social unrest. Korea was invaded and colonized by Japan in the early 20th century. In the middle of the 20th century, there were also large-scale wars and so on.

All this shows that, for a society, the supply of various resources outside the society and the thermodynamic balance of the whole society are very important to maintain the stability and prosperity of the society. Such situations in Japan, South Korea, and Vietnam have also occurred in other countries in Southeast Asia.

The interaction between social thermodynamic systems mainly includes the following aspects:

The first is the transfer of energy. That is to say, energy is transmitted from one social system to another, so as to satisfy the energy required for the balance of the other social system.

The second is the transmission of temperature. If the social temperature of one social thermodynamic system is high and the temperature of the other social thermodynamic system is low, heat will be transferred from the society with high social temperature to the society with low temperature. Finally, the balance between the two social temperatures is reached. This social temperature corresponds to the degree of prosperity of social technology and culture, and so on. That is to say, two adjacent societies will eventually share the achievements of these human civilizations. Of course, whether the other society can accept it depends on whether its social structure can accommodate these achievements of civilization.

The third is the mutual flow between basic society members. This mutual flow of basic society members is consistent with the mutual transfer effect of particles in a physical thermodynamic system. The mutual migration of basic society members is actually a process of information and energy transfer.

The fourth is the transfer of entropy. If two societies are in equilibrium, one of the societies has a higher entropy and the other has a lower entropy, then this entropy flow will pass from high to low. In a society with a low entropy, its entropy will increase, and then the total amount of information it has will also increase significantly.

Of course, entropy has another meaning, which is the degree of social order. For example, everything in a society is well-organized, the duties of managers are very clear, ordinary citizens also abide by laws and regulations, and the social order is very good. Then the entropy of such a society is relatively small. If the surrounding society is in disorder, the surrounding society will also be affected by the society with less entropy, so that the entropy of the adjacent society can also be maintained at a lower level. Of course, this involves a non-equilibrium, nonlinear problem. This nonlinear process requires a lot of energy to maintain. In addition, this entropy, which reflects the social order, is a kind of entropy on a more macro level, which is another meaning of entropy. It is different from the corresponding entropy of the total amount of information.

3.5 Interaction within different age groups

3.5.1 Human cognitive growth and interaction

On the one hand, different social interactions and growth environments lead to the emergence of different ethnic groups. Continuous social interaction within a nation can continuously strengthen the connection between basic society members, leading to the formation of a unique national society. In addition to considering the different social forms and structures due to different nationalities, we can also notice that each social individual has a growth process. At different growth stages of a person, his interaction with the whole society will also be very different. For example, when a person is in childhood, he mainly has social interaction with family members, and rarely interact with individuals from other families. By childhood, many children can enter kindergarten, and some even enter elementary school. At this time, it can contact other children more widely, so as to carry out interaction in a relatively simple society.

According to Piaget's division of children's developmental stages, children between the ages of three and seven are in the preoperational stage. At this stage, children already have a certain ability of representational thinking. However, there is still a big gap between the ability of expressive thinking and the ability of abstract thinking. Therefore, children can only carry out relatively simple thinking and reasoning. In this case, their interaction with other children is also relatively simple. It is mainly based on the appearance of the other party, as well as some visible and audible signals, etc. to communicate.

But by the teenage stage, they have been able to rely on concrete things for some simple logical reasoning. At this time, the content of their communication with other teenagers will become more complicated, and will use more symbols such as language to do it.

In youth, their abstract thinking skills are further developed. By this time, they have been able to carry out formal logical reasoning. Therefore, young people will analyze, summarize, synthesize and evaluate according to the causes and consequences of the people and events they have come into contact with, and the various facts they have obtained. With this ability, the knowledge and content they exchange with each other is also richer. At this time, youth began to involve some more abstract and higher-level civilization achievements accumulated by many human beings.

When they are about 20 years old, many young people have entered the university, and at this time they will determine their professional direction. Therefore, when communicating, usually only communicate with these individuals who are similar to their majors. The content of communication is more professional.

In adulthood, individuals in a society have to take on more family responsibilities. They would do more purposeful work in order to earn money to support their families and so on. At this stage, they can exchange more information with each other, mainly related to how to survive, how to protect their relatives and so on.

After reaching middle-aged, with the adulthood of children, a person's family burden begins to decrease. At this time, many middle-aged people begin to focus on how to promote the healthy growth of the next generation. At the same time, they will also think about their own growth process in the past. From the perspective of cognitive ability development, people's cognitive ability in this period has reached a very high level, so they think more deeply about various issues. At this stage, some middle-aged people can also do some work that they could not do before.

Therefore, the content of mutual communication between middle-aged people will naturally become more abstract. They will stand in a higher perspective to think about life's problems. They are also able to look critically at the causes of various things with a more objective attitude.

At the stage of old age, due to the limitation of physiological factors, a person's cognitive ability is constantly declining. In such a process of cognitive decline, it will be difficult for the elderly to think more deeply. But the elderly is able to sum up their past experiences very well. Therefore, at this stage, the elderly mainly exchanges various social experiences accumulated by himself/herself and conduct an objective reflection on his/her life.

It can be seen from the above analysis that during the growth of a person's life, his/her cognitive ability changes with age. A person's cognitive ability is also different at different ages. The content of his thinking is also different, of course, this also directly affects the social interaction between individuals of a certain age. And this impact will directly lead to changes in the entire social structure and form.

People of different ages communicate with each other in different ways and content, which also shows that there are many subsystems in a social system. If these subsystems are divided according to age stage and cognitive ability, they can be divided into children's subsystems, teenagers' subsystems, youth subsystems, middle-aged subsystems, and elderly subsystems. And if we consider the relationship between human physiological growth and gender interaction is very close, we can divide the whole society into three subsystems according to the strength of gender interaction. The first of these is the children subsystem that basically has no gender interaction. The second is the adult subsystems with stronger gender interaction. The third is the elderly subsystems where gender interaction begins to decline.

3.5.2 Children interaction

For children, because there is no or little gender interaction among their individuals, the children's subsystem is mainly social interaction related to information transmission. Children mainly carry out this kind of social interaction by exchanging various kinds of information, so that they can form a relatively close social connection between them. Finally, the social subsystem of children and adolescents is formed.

Various thermodynamic parameters of the social system can also be used to describe the children's social subsystem. These parameters include social pressure, social space, social temperature, entropy, internal energy, chemical potential, Gibbs free energy, and enthalpy, among others. When dealing with the relationship between a subsystem and the entire social system, we can regard other subsystems of the society as external environmental factors. So the children subsystem is actually an open subsystem. In this open subsystem, there are constant exchanges of matter, energy, social temperature, and social individuals. Of course, this social subsystem will also be under pressure from the whole society. Parameters such as social pressure, temperature, and space can all be described by the equation of state of an ideal gas. Of course, considering the interaction within the children's subsystem, we can also use the Van der Waals equation to describe this subsystem.

The pressure on the children's social subsystem mainly comes from the whole society. These pressures include various legal provisions to promote the development of children and adolescents. In addition, these pressures also come from the various demands placed on them by the family and society. When children generally enter schools for compulsory education, the school system will pass on various social pressures to them.

Generally speaking, the physical social space of the children's subsystem is basically the same as the physical social space of this social system. After all, where adults can go, children can basically go. However, considering that countries around the world have various regulations on the growth of children and adolescents. For example, access to some information will have some special restrictions on children. Such information access restrictions, as well as access restrictions on a small number of physical spaces, will compress the social space of children's social subsystems to a certain extent.

The temperature of the children's social subsystem is mainly related to the internal energy and entropy of the system. If it is a society with sufficient energy supply, the internal energy of children's subsystems will be relatively large. On the other hand, if the amount of information received by the children is very large, the entropy of the whole children's social subsystem will also become very large. Finally, the joint promotion of several factors has led to a significant increase in the temperature of the social subsystem of children.

Then we analyze the equation of state or Van der Waals equation, we can see that for the social subsystem of children, the change of parameters will produce various observable effects. For example, if the social space of the children's space remains unchanged, when the entire subsystem is under greater social pressure, the temperature of the children's social subsystem will be significantly increased. This means that the internal energy of the children's social subsystem will increase rapidly.

For another example, if the social interaction between children and children is enhanced, it is easier for children to exchange various information with each other. In this case, social interaction in the children's social subsystem will be enhanced. According to the Van der Waals equation, we can see that such a subsystem is more likely to form a condensed state. The formation of this condensed state is mainly related to temperature and pressure. If the pressure decreases or the temperature decreases, the subsystem is prone to a phase transition, from a relatively loose state to a relatively stable condensed state. The formation of this condensed state is often very confusing to many adults. From the perspective of adults, these children seem to know a lot of things, but when they put it into practice, they find that the behavior of these children is too naive.

3.5.3 Case study: Academic differences of students in the background of Chinese and Western cultures

The differences in how children in China and the West behave in their learning are an issue of interest to many. Usually people think that Chinese students seem to have high IQs and strong learning ability, and can easily grasp various knowledge. This is of course a prejudice, which makes it easy for teachers to ignore the tutoring of Chinese students. However, if we analyze the differences between Chinese and Western cultures from the perspective of Social Thermodynamics, we can actually find some regularities in them.

First of all, we regard Chinese students and Western students as two different social subsystems, so the various thermodynamic parameters of these two different social subsystems will naturally be different.

Due to the limitations of traditional Chinese culture, Chinese students may have much less social space than Western students.

From the perspective of social pressure, due to the constraints of various ethical and moral standards formed by traditional Chinese culture, the requirements for students are quite high. These high requirements are mainly reflected in requiring students to be filial to their parents at home, to respect teachers at school, and to be able to live in harmony with others in society. This all-round requirement puts great social pressure on students. In this respect, Western students are more advocating for freedom and the development of individuality. Therefore, the social pressure on Chinese students is far greater than that of Western students. Although the social space of Western students is larger than that of Chinese students, Chinese students bear greater social pressure, so from the perspective of the ideal gas equation of state, the result is an increase in social temperature. This shows that the temperature of the social subsystem of Chinese students is significantly higher than that of Western students. This also shows that the internal energy of Chinese students is much higher than that of Western students in their teenage years. This increase in internal energy usually means that more energy is expended in the brain's neural network than in Western students. From the point of view of non-equilibrium thermodynamics, the greater the energy consumed by the brain neural network, the greater the amount of information that the brain neural network needs to process. In fact, this result means that, if we only look at the subsystem of students in school, the knowledge that Chinese students need to master may exceed that of Western students. This is also why after many Chinese students go to Western schools, Westerners feel that Chinese students are more diligent and smarter.

But on the other hand, we can also notice that for Chinese students, the social interaction among their family members is relatively strong due to the more frequent information exchanges between them and their family members. This leads to a relatively weak social interaction related to information exchange between Chinese students and children from other families. Of course, this is also due to the pressure generated by the traditional Chinese ethical and moral culture. In contrast, the interaction between students from different families in Europe and the United States will be stronger. From the Van der Waals equation, this stronger interaction between individuals will lead to the easier formation of a condensed state in the social subsystem. This shows that compared with European and American students, Chinese students are less likely to form a condensed state. The relationship between Chinese students is relatively loose. The actual reflection is that the cooperation ability of European and American students is obviously much stronger than that of Chinese students.

The ability to cooperate is the foundation of scientific research. As Newton said, his results stood on the shoulders of giants. Therefore, Western students are good at getting inspiration from other people's thoughts when they are engaged in scientific research. Chinese students, on the other hand, are more inclined to complete a complex job from scratch. As a result, many students under the influence of traditional Chinese culture may have much weaker scientific research ability than Western students in their adulthood.

This issue has attracted the attention of many East Asian countries with strong traditional culture. For example, Japan has carried out reforms in education that are modeled on Western education, focusing on cultivating students' cooperative ability and personality development (Bjork & Tsuneyoshi, 2005; Bjork, 2011). Of course, it also retains many of the advantages of East Asian culture. This kind of educational exploration that attaches importance to students' cooperative ability has achieved obvious results in Japan. For example, since the beginning of the 21st century, Japan has basically received Nobel Prize winners in science every year.

This well reflects that it is equally feasible to educate and cultivate students' scientific research ability and cooperation ability in the traditional cultural circle of East Asia.

After quantifying some important social thermodynamic parameters, Section 9.3.3 will conduct a more detailed quantitative analysis of the academic differences between students in this Chinese and Western cultural context.

3.5.4 Adult group interaction

There is a big difference between the social subsystem composed of adults and the social subsystem composed of children. This difference is reflected in the fact that, on the one hand, adults have stronger cognitive abilities than children, and on the other hand, there is gender interaction among adults. The enhancement of cognitive ability means that the information exchanged between them is very complex, and the amount of information is relatively large. Therefore, the intensity of interaction between adults is also much larger than that between children and adolescents.

The existence of gender interaction means that in the adult social subsystem, the basic society members are mainly gender molecules or nucleus. That is to say, two adults of different genders are usually combined to form a whole. This whole is gender molecular or gender nucleus. There is also interaction between gender molecules or nucleus. This kind of interaction is very different from the interaction between individuals

In addition, the complexity of the adult social system is also reflected in that it does not mean that all the opposite sex can form gender molecules or nucleus with each other. In the adult social subsystem, there is also a single gender individual outside the gender molecule. The existence of these single gender individuals leads to the existence of gender interaction in the entire social subsystem, and affects the social interaction among other social members.

Two different interactions coexist in the adult subsystem, which may lead to further splitting of the adult social subsystem into two different social subsystems. There is interaction between these two different social subsystems.

However, the existence of gender molecules and single gender individuals in adult social subsystems may also lead to some interesting thermodynamic phenomena in society as a whole. If the social temperature is relatively high, then the two basic society members can exist in the adult social subsystem at the same time. But if social temperature or social pressure changes, basic society members may transform into each other. For example, if the social temperature is relatively low, then a single gender individual will easily combine with other individuals of the opposite sex to form gender molecules. If the social temperature is high enough, the internal energy of the whole society will be relatively large, which will easily lead to stronger social interaction than gender interaction, which will cause the molecular bond between gender molecules or nucleus to break, resulting in a rapid increase in the number of single gender individuals in the entire society. This can be seen from the comparison between ancient China and modern information society.

In the ancient Chinese traditional society, because the ability of the whole society to obtain energy is relatively weak, the temperature of the whole society is relatively low, so the combination of gender molecules is relatively close. So in society as a whole, there are very few single people. In the modern information society and the industrial revolution, human beings can more actively obtain energy from the environment, which leads to the rapid growth of human society's internal energy and naturally high social temperature. This very high social temperature can easily cause the breakage of the gender molecular bond, resulting in more single people being reluctant to form a family. Or even it is already formed a family, and it is easy to get divorced and so on.

If gender interaction is not considered and only social interaction is considered, the thermodynamic properties of adult social subsystems are basically similar to those of other social subsystems, and they can be combined with ideal through changes in parameters such as social pressure, social temperature and social space. The gas equation of state and the Van der Waals equation are used for analysis.

3.5.5 Interaction of the elderly

Unlike children and adults, the cognitive abilities of the elderly, especially those over 80 years old, are in decline. Although relying on the accumulated experience in the past, the elderly can still deal with some more complex cognitive problems, but when dealing with new problems, they may be unable to solve them due to the limitation of cognitive ability.

In addition, the gender interaction of the elderly will become weaker and weaker. Therefore, the interaction between the elderly is mainly reflected in the social interaction related to information. The relative enhancement of this social interaction also leads to an increase in the frequency of mutual connections among the elderly.

Although many elderly people, like children, have weak gender interactions. However, the life experience of the elderly is very rich, and they have accumulated a lot of valuable life experience. In addition, many elderly people are also professionals in a certain area when they are young, and the content they communicate will be more professional. Therefore, the social interaction of the elderly is more diverse, with high complexity and professionalism. The natural structure of the social subsystems of the elderly is also more complex, diverse and professional.

In addition, many elderly people have partnered with the opposite sex when they are young, forming a gender molecular or gender nucleus. Then the gender molecule or gender nucleus becomes stronger and more integrated in old age. Therefore, the social interaction in the elderly population is mainly the interaction of the gender molecular or gender nucleus. In the event of divorce or widowhood, separate gender individuals will reappear. However, considering that physiologically, the gender interaction of the elderly is already very weak, so the gender interaction in the social subsystem of the elderly is very weak. This also avoids various problems that may arise in the single group due to the temperature change of the large social system in the adult social subsystem.

For a normal society, the proportion of the elderly in the whole society is relatively small. Especially in some traditional family-oriented societies, the elderly will spend their later years in the family. In this case, the elderly mainly communicate closely with their family members in the family, and the frequency of communication between the elderly in other families will gradually decrease. This also means that in a society without the problem of aging, although the social interaction related to information among some elderly people will be enhanced compared to when they were young, in different social types, the differences of social interaction among the elderly are still very large. Especially in family-oriented society, the social interaction of the elderly with the elderly in other families may decline due to the inconvenience of mobility. According to the Van der Waals equation, such a subsystem does not easily form a condensed state. Naturally, it will not become a social problem.

At present, with the advent of the so-called aging society, the proportion of the elderly in the entire social system is increasing. With the help of modern information technology, especially the Internet communication technology, the elderly with limited mobility can also fully communicate with old friends thousands of miles away while sitting at home. This means that on the one hand, the aging social subsystem will become larger and larger in some aging societies. On the other hand, social interaction among the elderly is also becoming more and more intense. Therefore, when the social temperature changes, the aging society is prone to phase transition. If the social temperature decreases to a certain extent, or the intensity of social interaction becomes stronger, a condensed state may be formed. The entire social subsystem of the elderly will become more organized, and the goals and demands of actions may become clearer. For example, with regard to the demands of some social issues, the views on the we-media where the elderly occupy the majority may be completely opposite to the views on the we-media where the young people occupy the majority.

References

Bjork, C. (2011). Imagining Japan's 'relaxed education' curriculum: Continuity or change. Reimagining Japanese Education: Borders, transfers, circulations, and the comparative. Symposium Books, 147–169.

Bjork, C., & Tsuneyoshi, R. (2005). Education reform in Japan: Competing visions for the future. *Phi Delta Kappan*, 86(8), 619–626.

Hinton, K., Baliga, J., Feng, M., Ayre, R., & Tucker, R. S. (2011). Power consumption and energy efficiency in the internet. *IEEE Network*, 25(2), 6–12.

Hollingsworth, J. Seo, Y. & Bae, G. (2020). South Korean authorities encourage men to marry foreign women. But their brides often become victims of abuse. https://edition.cnn.com/2020/08/02/asia/foreign-wives-south-korea-intl-hnk-dst/index.html.

Japan Data. (2018). "2016 Vital Statistics" report issued by the Ministry of Health, Labor, and Welfare. https://www.nippon.com/en/features/h00174/

Mbizah, M. M., Valeix, M., Macdonald, D. W., & Loveridge, A. J. (2019). Applying the resource dispersion hypothesis to a fission–fusion society: A case study of the African lion (Panthera leo). *Ecology and Evolution*, 9(16), 9111–9119.

Mech, L. D. (1999). Alpha status, dominance, and division of labor in wolf packs. *Canadian Journal of Zoology*, 77(8), 1196–1203.

Ministry of Civil Affairs of the People's Republic of China. (2020). China Civil Affairs Statistical Yearbook 2020. CD-ROM, Beijing: China Statistics Press.

Packer, C., & Pusey, A. E. (1997). Divided we fall: cooperation among lions. *Scientific American*, 276(5), 52–59.

4

Thermodynamic parameters of biological systems

4.1 Comparison of biological and thermodynamic systems

4.1.1 Energy input and output

The previous chapters have been qualitative discussions of the social thermodynamic system. If it is necessary to accurately predict the future development trend of the social thermodynamic system, a relatively accurate calculation process is required. In order to achieve this purpose, this chapter will start to define some parameters of social thermodynamic system more accurately. In this way, we can directly obtain the quantified parameters and use the existing formulas in physical thermodynamics to deduce and calculate.

Of course, such quantification is not easy given the complexity of the social thermodynamic system, and there are even many social problems that cannot be quantified. Especially when it comes to human knowledge, we cannot simply use a quantitative comparison to reflect which culture has a greater amount of information. For example, ancient

Chinese culture is difficult to make a simple quantitative comparison with the information carried by modern science and technology. On the one hand, the content contained is different, and on the other hand, it also involves a problem of knowledge inheritance.

Therefore, the quantification work started in this chapter is only to quantify some of the parameters as much as possible. And on this limited quantitative basis, it reflects some important macro trends of global human social changes.

First of all, in terms of energy input and output, the energy of a physical system can be directly described by quantifiable data. And the physical system is simpler, so to describe the energy of a moving object, only a few parameters are needed to achieve the purpose. For living systems, some biological organisms are themselves a very complex physical system. And such a biological system also needs to move as a whole, and the physical parameters it involves are naturally much more complicated than a single physical system.

Another problem is that biological systems have to deal with a large amount of information, which is itself very complex. The size of the amount of information is related to the number of states the system has and the difference in energy distribution. Therefore, even for the same physical system, if the energy distributions involved in the microscopic subsystems of these physical systems are different, the amount of information they have will be very different. How to link information and energy involves a very complex statistical physics problem.

For a simple example, even for physical Newtonian mechanics, the multi-body motion system is already very complicated, but we can still describe it with parameters such as the mass of several objects, the interaction force between them, and the speed of the object. But for a biological system, in addition to the overall quality and movement speed of the organism, the biological brain neural network system also thinks and judges. This process of thinking and judgment will also produce different responses with various external stimuli. This is the physical and psychological aspects of performance that belong to the biological individual. How to use physical formulas to describe the psychological performance of a biological individual, its specific behavior, the energy it carries and the energy it consumes, etc., which are relatively complex issues.

Of course, compared with the physical system, the energy input and output of the biological system is also very complex. Although we say that all energy comes from the sun, the energy input of a physical system is relatively simple in terms of specific energy input. It is usually the input and output of energy caused by various interactions. However, in addition to the daily changes in the environment that the biological system can perceive, it is also extremely complex in terms of the source of food. Although herbivores only need to eat plants to obtain energy, for some animals at the top of the food chain, the food they can eat is very large and very complex. For humans, in order to complete the requirements of work, humans will also obtain energy through water power, fire power, nuclear power, etc. These energies can meet the energy input of making food and obtain nutrition, and can also meet the energy needs of one's own travel and work, etc. Such complex energy requirements are completely incomparable to simple physical systems.

Directly linked to the input and output of energy is the transfer of heat. Heat is essentially energy. The main source of energy in biological societies is food. Different foods have different calories. These heats will eventually be converted into the biological kinetic energy of the individual and the energy required by the neural network system to process information.

In physical systems, heat is primarily derived from the difference in temperature between the two systems. If the temperature of one system is high and the temperature of the other system is low, then there will be heat transfer at this time. The same is true in biological systems. If a creature lacks enough energy to meet the needs of daily life, for example, a creature is in a state of starvation, then the creature must obtain external heat to meet its own physiological needs. The acquisition of these calories mainly comes from a variety of different foods, which is what we call the calories of food in our daily life. Under the condition of underdeveloped production technology, human beings generally lack enough calories. At this time, it is necessary for the whole society to increase the yield of crops, and social individuals to eat as much food as possible. However, with the rapid development of production technology, due to the abundant production of food, human beings have excessive calorie acquisition. Of course, when this calorie intake is excessive, it will lead to problems and diseases such as obesity and high blood pressure.

Social thermodynamics is the expectation that some important social parameters can be quantified into thermodynamic parameters that can be of the same dimension as the environmental physical thermodynamic system. In this way we can compare the social thermodynamic system with the environmental physical thermodynamic system. In this way, according to the 0th law of thermodynamics, if the temperature of the environment is higher than the social temperature of the social thermodynamic system, the social thermodynamic system will be able to easily obtain energy from the environment without consuming additional energy. And if the temperature of the social thermodynamic system is higher than the ambient temperature, the social thermodynamic system will not be able to obtain energy from the environment freely. However, by consuming additional energy, the social thermodynamic system can still obtain energy from the environment. This is the same as the cooling or heating effect of an air conditioner.

4.1.2 Relationship between various thermodynamic parameters

In terms of various thermodynamic parameters, there will be certain differences between the thermodynamic parameters of biological systems and those of physical systems. For example, in terms of volume, the volume of a physical system refers to the space occupied by matter. These substances can be gases, liquids or solids. This space can be measured using special measuring tools. On the one hand, biological systems occupy space like other substances to form volumes; on the other hand, because biological systems can also process various information, especially with the support of some advanced information technology, biological individuals can feel the existence of social space. They can feel that this social space will far exceed the size of the physical space in which they live. For example, in the virtual world of the Internet, if an individual plays online games in it, the size of the social space he feels is far from a very small range like the physical space where he lives.

This is the difference between social space and physical space. The composition of social space is more complex. It includes not only the physical space, but also the sense of space of the human psychological effect produced by the information transmission.

The same is true for the parameter pressure. For physical systems, the existence of various pressures is usually derived from electromagnetic interaction or gravitational interaction. For example, the existence of atmospheric pressure in the atmosphere is caused by the gravitational action of the earth. And some other pressures are essentially generated by electromagnetic interaction. For example, inside the engine, the pressure generated by the piston of the engine is generated by an electromagnetic interaction process such as gasoline combustion.

For biological systems, the pressure experienced by individual organisms will be significantly different. For example, for a biological individual living in a very small space, it will feel the pressure generated by the small space. This pressure is similar to physical pressure. But on the other hand, a biological individual may also feel the pressure of information transmission. For example, if a person hears a very sad news, his mental state will be relatively tense at this time, and he will feel very obvious mental pressure. Although this kind of mental pressure is produced by psychological aspects, it actually directly causes the actions of biological individuals. And the actions of such individual organisms will naturally lead to changes in the momentum of the individual organisms. From this point of view, mental pressure can actually produce observable physical effects. Prove that the biological individual is subjected to the action of force.

Therefore, when we deal with the thermodynamic problems of biological systems, we must also consider the existence of this information pressure. The social pressure on living things is actually composed of two aspects: physical pressure and information pressure.

The society composed of living things also has temperature. Similarly, this temperature, like pressure and social space, can also be divided into two aspects: social physiological temperature and social information temperature. From the perspective of social physiological temperature, it reflects the active situation of individuals in this social system. If it is in a biological society, the individuals in it are very active and the amount of exercise is very large, which means that the kinetic energy of the individuals in this biological society is very large. This is similar to the kinetic energy of the molecular in the gas, so to judge the social temperature of a biological society, we can make a comprehensive judgment from the quality of the biological individuals formed in this biological society, their degree of movement, and the speed of movement. If the kinetic energy of the biological individuals in a biological society is very large, it means that the physiological temperature of the biological society will be relatively high. For example, in the system composed of dinosaurs 300 million years ago, because some dinosaurs had very large bodies and many dinosaurs moved very fast, we said that these dinosaurs formed a biological social system, and its social physiological temperature is relatively high.

Of course, for biological systems, in addition to social and physiological temperature, we can also notice that if the amount of information transmission in a biological system is very large, and this amount of information represents the neural network system in this biological system. The amount of the number of states, the more the number of states, the greater the entropy of the entire system. The greater the information entropy corresponds to another kind of temperature increase. Therefore, in addition to the physiological temperature, the social temperature of the biological social system also includes the information temperature. This information temperature is directly related to the amount of information in a biological society. Generally, the biological society composed of creatures with higher levels has a larger amount of information, so its information temperature will naturally be very high.

4.1.3 Information flow

At the level of human society, due to the high level of civilization of human society, the human brain is also very complex, and the neural network function is also very perfect, so the amount of information in human society is much higher than the total amount of information in animal society.

The existence of information also means that there is a flow of information. If the information flow rate can be measured, then we can obtain the information flow parameter by multiplying the amount of information by the information flow rate.

The existence of complex information flow is an important feature that distinguishes human society from animal society. Although there are also information flows in animal society, the direction and way of these information flows are very simple. For example, the information flow in a society composed of bees, when a bee finds that there is nectar available, it will pass this information to other bees. The transmission of this information often takes the form of hormone volatilization.

But for human society, the flow of information transmitted by humans is obviously much more complicated. Because humans can use language, of course, they can also transmit information in various ways such as sound, vision, etc., so the content of the information flow transmitted by humans is not just a simple information as to whether there is nectar in a certain place to other individuals.

The transmission of Information flow includes broadcast method and point-to-point method. The method of broadcasting is usually the broadcasting of a message to multiple individuals at the same time. In the absence of information technology support, it is usually achieved by means of assemblies. Through the gathering of crowds, one person can transmit his own information to many other individuals at the same time. After the use of information technology, an individual can spread to other more individuals through radio broadcasting, television, the Internet and so on. With the help of information technology, the number of individuals receiving information can reach hundreds of millions or even billions.

The flow of information flow also adopts a point-to-point approach. The traditional method is used to carry out point-to-point communication, which is generally the direct transmission of information through language. After the advent of telephone technology, two individuals at different distances can be connected together by a line through a switch. At this time, point-to-point communication can be achieved between the two individuals. With the use of Internet technology, such point-to-point communication becomes more efficient.

Whether it is broadcasting or point-to-point information dissemination, it involves the flow of information flow. We can also notice an important difference between the flow of information and the flow of matter. The flow of matter must follow the law of conservation of particle number, while the propagation of information flow can not obey this conservation of particle number. Because of the way of broadcasting, a small piece of information can be spread to countless individuals at the same time. Of course, this does not violate the law of conservation of energy, because each piece of information actually consumes energy in the process of copying.

4.2 Energy production and consumption in biological systems

4.2.1 Levels of energy generation

The most basic energy provided to a social system is solar energy. Through solar energy, plants and animals on earth can grow. At present, the widely used oil in human society is actually because the ancient plants on the earth received solar energy to grow greatly. And buried deep in the ground to become fossils and eventually oil. The nuclear fusion energy that may be used on Earth in the future is also derived from isotopic elements of hydrogen and helium released by the sun.

Of course, the energy directly supplied to human society is more diversified in form. In ancient times when science and technology had not been developed, the main energy sources used by human beings included rice, wheat, fruits, vegetables and other foods grown through solar photosynthesis. One level up, there are herbivores that eat plants, and then one level up are carnivores. The food energy obtained by humans comes from plants, animals, etc., and is at the top of the food chain.

However, in addition to the food needed for human physiological growth, ancient human societies could also obtain energy through the flow of water and the atmosphere. Through this energy, food can be deeply processed, so that humans can obtain animal and plant energy more efficiently. By burning plants, humans can obtain the energy of fire to meet the needs of cooking food and lighting. This is also a basic way to improve the energy acquisition efficiency of animals and plants. At the same time, using fire, humans can also transmit information and improve the efficiency of information acquisition. It can also raise the physical temperature of the environment and help humans resist the cold. The discovery of fire is therefore the starting point for human society to utilize energy at a higher level. The use of energy can not only improve the quantity and efficiency of food acquisition, but more importantly, the use of energy promotes the quantity and efficiency of information transmission. Of course, it also further promotes human social interaction. It has laid a solid foundation for the development of human society.

However, the use of various energy sources also brings a series of problems to human society. These problems include environmental pollution, global warming and so on. Therefore, while maximizing the use of energy to promote the development of human society, human beings are also paying attention to the improvement of energy utilization efficiency. Among them, the use of various new energy sources is becoming more and more extensive.

In the composition of new energy, the most important way is to use solar energy directly. Such utilization methods include the use of solar photovoltaic power generation facilities to generate electrical energy and so on. And wind and hydro energy have also been widely used.

Of course, the cleanest energy source is nuclear fusion energy. This way of nuclear fusion is the most important way the sun radiates energy and matter. In fact, only a small part of the nuclear fusion material needs to be used on the earth to obtain energy that meets the needs of global use, and basically there will be no problems of environmental pollution.

All these different levels of energy, input into the social system composed of humans or other animals, constitute the most basic energy source of the entire biothermodynamic system. Different biological social thermodynamic systems have different levels of efficiency when using energy. Thanks to the long-term biological evolution, the biological efficiency of biological energy utilization is very high. For example, green plants use solar energy for photosynthesis, and their efficiency can reach a very high level. While humans use industrial technology to convert solar energy into electrical energy, the energy conversion efficiency is much lower. In terms of obtaining energy through food, the physiological systems of various organisms are also extremely efficient at converting energy.

However, with the advancement of technology, the industrial system of human society is also rapidly improving the efficiency of energy utilization, thus laying a solid foundation for the more harmonious development of human society. This improvement in energy utilization efficiency is critical to maintaining the normal operation of the entire biothermodynamic system from a thermodynamic point of view.

4.2.2 Composition of social energy

Like other ecosystems, the social system is a nonlinear system that is far from equilibrium and is maintained by the input and output of energy. Such nonlinear systems have a high degree of complexity.

For a social system, the energy composition mainly includes:

$$E = U + B \tag{4-1}$$

Among them, E is the total energy of the social system; U is the internal energy of the system; B is the external energy. These external energies are mainly solar energy, the energy provided by the nutrition of the food, the energy generated by the artificial power system, etc. Use the differential operator d to represent the change in energy. The unit of energy is the joule, or J for short.

Of course, when we deal with thermodynamic problems, one of the problems we often encounter is the change of energy, so we often use the differential form, namely

$$dE = dU + dB \tag{4-2}$$

Because a social thermodynamic system is actually built on a physical thermodynamic system, the energy of the whole system can actually be divided into physical energy and social thermodynamic energy. This is mainly manifested in the internal energy U.

We refer to the internal energy of physical thermodynamics as U_p, and the energy generated by the social activities of the social individual itself is called social internal energy. These social activities include their own nutritional needs and physiological needs and so on. The internal energy is represented by U_s, then the thermodynamic internal energy of the whole society can be expressed as:

$$U = U_p + U_s \tag{4-3}$$

Under normal circumstances, since the temperature of the human body is constant, the physical internal energy U_p is usually constant. In this way, when solving the change of internal energy, it can be considered that:

$$dU_p = 0 \tag{4-4}$$

Then

$$dU = dU_S \tag{4-5}$$

That is to say, in general, changes in internal energy usually reflect changes in social internal energy.

Another concept closely associated with energy is heat. Calories are more commonly used in everyday life. For example, the number of calories in a certain food is very important for the normal operation of a physiological system. The unit of heat is the calorie, an auxiliary unit of energy, which is consistent with the "joule" in the International System of Units. 1 calorie = 4.184 joules.

So the change in heat is actually the change in energy. If the energy changes, it is absorbed or radiated as heat. Which is

$$Q = E \tag{4-6}$$

or

$$dQ = dE \tag{4-7}$$

where Q represents heat.

Math further

If it is considered that the change of entropy is only affected by the change of energy, then:

$$dS_h = \frac{\partial S_h}{\partial E} dE = \frac{1}{T} dQ \tag{4-8}$$

This is due to changes in entropy caused by changes in energy. We may call this entropy a thermal entropy.

There is another entropy in the social system, that is, changes in the amount of information lead to changes in the entropy of the entire society. This is called information entropy, denoted by H.

$$dH = \frac{1}{T} dQ_H \tag{4-9}$$

The calculation of the amount of information is complicated because of meaningful information. Therefore, this information entropy can be directly linked to the energy consumed by the neural network of the brain. This is considering that all information processing goes through the neural network of the brain. In this way, according to the law of conservation of energy, the energy consumed by these informations is the energy consumed by the neural network of the brain. Conversely, we can reduce the size of the amount of information according to the energy consumed by the neural network through formula (4-9).

Another important concept related to heat is the concept of heat capacity. It represents the change in the system's internal energy caused by a unit temperature change. The formula for heat capacity:

$$C_v = \frac{dQ}{dT} = \frac{dU}{dT} \tag{4-10}$$

Because the energy used by the social thermodynamic system and the physical thermodynamic system is essentially the same. Therefore, various forms of energy in the physical thermodynamic system can be directly transplanted. That is, when analyzing the social thermodynamic system, we do not need to create some new forms of energy.

4.2.3 Case study: Comparison of energy consumption between dinosaur society and human society

In terms of individual size, humans are much smaller than dinosaurs. It seems that humans consume far less energy than dinosaurs. However, the energy consumption of a society is composed of many factors. Here we can comprehensively analyze the social energy consumption of two different biological compositions.

For human society, based on the average human body weight of 70 kg and about 1 kg of food (or equivalent to 10 kg of grass or leaves) a day, it is obviously impossible to compare with a dinosaur. However, in addition to the physiological energy consumption of human beings, a person also needs to live in a house, drive a car, etc., which is impossible for other animals such as dinosaurs to enjoy. Therefore, these other factors must be considered when estimating, which can be roughly classified as follows:

1. The comparison of the mass

(1) The house

If one lives in an apartment in an urban area, the area is about 80 square meters, with a reinforced concrete frame structure. The floor thickness is 0.3 meters, and the ground volume is 24 cubic meters. The wall is 3 meters high and 0.2 meters thick. Excluding the wall shared with the next-door neighbor, the wall volume is about 18 cubic meters. Therefore, the total amount of reinforced concrete for the walls and the ground is about 42 cubic meters. Calculated according to the specific gravity of reinforced concrete of 2500 kg/m^3, the total mass of the house is 105,000 kg. Considering the average family of three people, the average weight of each person to the house is 35,000 kg. This is more than the mass of many large dinosaurs.

(2) Vehicles

In many social systems today, humans will own cars. Although compared to the house, the mass is not large, about 1,100 kg. The average is only more than 300 kg per person.

However, consider the public transport system throughout the city. Such as buses, subways, etc., these tools will also add a lot of average weight to humans. For example, there are 15,000 buses in a city, each with a mass of 8,000 kg; about 100 subway trains, each with a mass of 240,000 kg. The city has 13 million resident and floating population, and the average weight gain per person is about 12 kg.

Of course, there are buildings in subway stations. Each subway station has a construction area of about 30,000 square meters. All of them are made of reinforced concrete, and their mass reaches 50,000,000 kg. The total number of subway stations is about 100, and their total mass: 5,000,000,000 kg. In addition, the tunnel construction and rails should be considered. About 10,000 kg/m, a total of $230 \times 2km$, the total weight of the tunnel is: 5,000,000 kg. Average to each citizen, about 400 kg.

For cars, roads, interchanges and tunnels are also involved. If the total length of the city road is about 500 km, about 100,000 kg/m, the total weight reaches 50,000,000,000 kg. The average weight gain per person is 4000 kg.

(3) Other

Including the construction facilities, office buildings, commercial buildings, etc. required by factories that produce daily necessities, these are difficult to estimate because they are distributed in different regions of the country. However, due to the large number, it is estimated that on average everyone will add a lot of weight. It should be roughly the same as traffic.

That is to say, combining the above three items, the mass occupied by one person is about 45,000 kg. Comparable to the heaviest dinosaurs in the age of dinosaurs, or the equivalent of at least three meat-eating Tyrannosaurus rex.

2. The comparison of energy consumption

Assuming that dinosaurs eat 200 kg of plants per day, and each kilogram of plants contains 150 kcal of energy, each dinosaur consumes 30,000 kcal of energy per day. In comparison, the daily energy consumption of human nutrition is about 1500 kcal. In addition to this, there are also the following aspects of energy consumption.

(1) Car consumption

Suppose a car consumes about 4 liters of fuel per day, which is equivalent to 40,000 kcal of energy. This is just fuel that has been refined. The fact is, from oil extraction to refining and processing, to transportation to gas stations, etc., the efficiency of each link cannot reach 100%, and each link consumes extra energy. Therefore, efficiency must also be considered. Calculated at 50% efficiency, the energy consumption is 80,000 kcal.

(2) Electricity, gas, etc.

If a household consumes an average of 150 kWh of electricity per month and 3 kWh of electricity per day, it is equivalent to 2,700 kcal. If the power generation efficiency is 50%, and the coal mining and transportation efficiency is 50%, it is equivalent to consuming 10,000 kcal of energy.

The monthly consumption of natural gas is about 20 kg, which is equivalent to 0.6 kilograms per day, and the energy consumption is 10,000 kcal. The energy consumption efficiency is relatively high and is calculated by 70%, which is equivalent to the actual energy consumption of 15,000 kcal.

(3) Energy consumed by public transport, house, clothing, food processing, etc.

According to the previous calculation of mass increase, it can be roughly estimated that the energy consumed by this part can be calculated by the ratio of 1/4 of the energy consumed by itself, that is, the increased energy consumption is about 30,000 kcal.

The total energy consumed by such a family of three is: 135,000 kcal energy a day, and the average energy consumption per person is 40,000 kcal energy.

This is also more than the energy consumed by a large herbivorous dinosaur.

3. Conclusion

From the rough estimates above, humans are of course much smaller than dinosaurs, but in terms of resources consumed by humans, the average energy consumption of one person has exceeded the resources consumed by a large dinosaur. This is basically equivalent to the resources consumed by the three Tyrannosaurus Rex.

Of course, if the global population can live a life like that of the Americans, it is estimated that human society can consume more resources and energy.

4.3 Social space

4.3.1 The volume of matter and biological social space

The volume occupied by a substance is very intuitive. You can directly multiply the base area S by the height h to get the physical volume occupied by the substance. It is expressed by the formula

$$V_p = Sh \tag{4-11}$$

For biological society, we can also use a similar volume calculation method to calculate the physical space occupied by these creatures. It's just that unlike gases, liquids, and other substances, the physical space occupied by biological societies is mainly distributed on the surface of the earth. So if viewed from far away, these creatures appear to be living on a flat surface. But if we take into account that these biological societies still have a certain height, we can view the physical space occupied by these creatures as a disk shape. It is assumed that the height of the physical space occupied by the biological society is fixed, so its physical volume is mainly related to the area S of the disc.

But living things are different from inanimate matter, and information is exchanged between living things. The biological brain also actively thinks. Therefore, looking at a creature in terms of information transmission, it also possesses something called information space. For a creature, the more complex the information he processes, the larger his information space will be. Therefore, the higher the level of the animal, the greater the information space it has.

For example, although a person confines himself to a very small room, the physical space occupied is of course very small. But if this person uses some technologies of virtual reality, and then uses these technologies to play online games, we can find that the time and space that this person actually feels is very large. This is his information space. In his information space, this game player can gallop in time and space with a radius of hundreds of thousands of kilometers. We use V_H to represent the information space here. Then the size of social space can be expressed as

$$V = V_p + V_H \tag{4-12}$$

At present, based on the blockchain of the Internet, an information technology called metaverse can be constructed (Heuser, 2003). In this metaverse, the information space constructed by information technology is very vividly linked with the physical space. Therefore, although a person's physical space is very small, if he is playing games or dealing with work-related tasks in the metaverse, his information space in the metaverse is very large.

On the other hand, we have emphasized many times before that information technology can influence the intensity of social interaction. That is to say, the functions of information technology are multi-faceted. In social thermodynamics, we believe that information technology has two functions: constructing social space and changing social interaction.

That is to say, if the use of information technology is increased, on the one hand, social space can be expanded, and on the other hand, the intensity of social interaction can be changed. And this change may be to enhance the social interaction coupling coefficient, or it may weaken the social interaction coupling coefficient. The key is how information technology is used.

However, unlike the physical space, the information space is more abstract, and how to measure the information space is one of the most important requirements to achieve computable purposes. Otherwise, we can only perform qualitative analysis of the information space.

In order to be able to perform calculations, the information space needs to be given a unit of volume. According to the International System of Units, the unit of this volume is m^3.

The proportion of information technology to increase social space can be estimated according to the information technology to improve the efficiency of information transmission. For example, the increment of living space caused by information technology can be expressed as:

$$V' = \alpha V \tag{4-13}$$

Where V is the physical space, and the V' is the information space. The coefficient α is used here to represent the increment multiplier caused by information technology.

From here, we can get an inspiration, that is, after introducing an information technology, the effective diameter of an individual can be increased. Therefore, it is assumed that after the introduction of a new information technology, the effective radius of the individual is increased to r. Therefore, the effective space occupied by him is

$$V = \pi r^2 h \tag{4-14}$$

Here, it is assumed that the individual's social space is a cylinder with a height of h. If it is a cylinder, the radius of the base is r. If there are n individuals, the total social space will be:

$$V = n\pi r^2 h \tag{4-15}$$

Assuming that without using any information technology, the effective radius of individuals communicating through their own physiological functions is R, and the physical space is V_p

Then, without using any information technology, the volume of the entire social space is:

$$V = V_p + \pi n R^2 h \tag{4-16}$$

After using information technology:

$$V' = V_p + \pi n r^2 h \tag{4-17}$$

Explain that the increment of social space after the use of information technology is:

$$\Delta V = V' - V = \pi n \left(r^2 - R^2\right) \tag{4-18}$$

To confirm the information technology increment coefficient α in the formula (4–13), it can be calculated according to the attributes of information technology, or it can be measured according to the actual data.

For information technology that can communicate over long distances, such as radio, we can use the collision cross section in physics to analyze it. With information technology, the cross-section of interaction between two social individuals will increase significantly, that is, interaction between two individuals is more likely to occur. The calculation formula of the collision section of the individual is:

$$\sigma = \pi d^2 \tag{4-19}$$

where d is the effective diameter of the individual. If information technology is used, this effective diameter can be significantly increased. In physics, the relationship between the collision frequency of an individual and the collision cross section can be expressed as:

$$p = nf(T)\sigma = \pi n f(T) d^2 \tag{4-20}$$

where $f(T)$ is a function related to social temperature, and n is the density of individuals in the society.

If the collision frequency between individuals in a society that does not use any information technology is used as a benchmark, the collision cross-section of an individual can be obtained by dividing the collision frequency of individuals after using statistical information technology by the reference frequency. The effective radius r occupied by the individual in the information space after the information technology is used can be obtained from the diameter of the collision section, so as to calculate the size of the information space. Which is:

$$r = \sqrt{\frac{4p}{\pi n f(T)}} \tag{4-21}$$

If there is no information technology, using the corresponding collision frequency as the reference frequency p_0, it can be calculated

$$R = \sqrt{\frac{4p_0}{\pi n f(T)}} \tag{4-22}$$

Now we can estimate some of the constants involved in Equation (4–22) with a loudspeaker device. Assuming that within a radius of one meter, the frequency at which the information transmitted between two individuals can be received by each other is p, so the frequency of this collision can be set to p. Considering that the attenuation of sound is an inverse square law, as the radius increases, the collision frequency between two individuals will become smaller and smaller, and it will attenuate inversely proportional to the square of the distance.

Typical information technology includes point-to-point communication tools – telephone, broadcast communication tools – TV, face-to-face communication tools – the Internet.

If it is point-to-point communication, although the communication distance is longer, it is only suitable for the enhancement of interaction between individuals. This enhancement did not significantly increase interaction frequency. Although broadcast-type information technology can realize point-to-surface broadcasting, it enhances the frequency of interaction between a certain individual and some other individuals. But it did not significantly increase the frequency of interaction among all individuals. The face-to-face Internet can avoid the above-mentioned two information technology problems and realize face-to-face information dissemination, thereby significantly enhancing the frequency of interaction between individuals.

According to the above analysis, it can also be seen that although information technology has increased the frequency of contact between individuals in society and other individuals. But if there is only contact and no communication, there will be no increase in social interaction between social individuals. Even if two individuals come into contact with each other through information technology, but instead hate or deceive, information technology may also weaken social interaction.

4.3.2 Case study: Different versions of the Internet improve social space

After the use of network technology, due to the increase of information space, the frequency of contact between social individuals will increase. The Web is a very important application in the Internet (McPherson, 2009). It can significantly enhance people's ability to access information, while also facilitating communication between individuals in societies around the world. If Web1.0 is used, the contact frequency will be increased to p, so that the collision frequency between social individuals of this version of the web can be calculated.

Because Web1.0 mainly allows various companies, including small and medium-sized enterprises, to display their work through the website. Therefore, in terms of increasing the frequency of individual collisions in society, it is mainly to provide a channel for mutual communication between ordinary people and professionals. For a society with a population of 100 million, the number of registered business entities can reach more than 10 million. Of course, there are far fewer business entities that can match the profession and hobbies of a certain social individual. We consider that what a person needs to face in life is mainly education, medical, commercial services and other products, and there are only a few hundred that need to be contacted.

After using Web2.0 technology (O'reilly, 2009). It means that ordinary people can also express their opinions directly on the Internet. These Web2.0 networks include blogs, Twitter, Facebook, and more. This enables direct point-to-point connections between ordinary people. Of course, social individuals have different professional levels. For a very professional field such as scientific research, the number of people who can engage in such professional research is of course very small. However, compared with Web1.0 or the era of relying only on traditional libraries to access materials, these scientific researchers can still contact more professional scientific researchers around the world in Web2.0 technology.

Of course, for ordinary people, in the technology of Web2.0, the possibility of finding social individuals who have similar views and can continue to communicate will increase to several thousand.

If a person's social interaction can only be limited to a village of several hundred people without using any information technology, then the frequency of collision between one person and others is about $p_0 = 30$. Then we assume that without the use of information technology, the size of the radius of a person's information space is 1 meter. So

$$1 = \sqrt{\frac{120}{\pi n f(T)}} \tag{4-23}$$

You can get:

$$nf(T) = \frac{120}{\pi} = 38.2 \tag{4-24}$$

According to the above estimation, we can set the collision frequency of social individuals as $p = 300$ under the Web1.0 technical conditions. Then it can be calculated that under the technical conditions of Web1.0, the size of the ground radius of the information space of a single social individual is:

$$r = \sqrt{\frac{4p}{\pi n f(T)}} = \sqrt{\frac{1200}{38.2\pi}} \approx 3.2(m) \tag{4-25}$$

The collision frequency of Web2.0 is ten times that of Web1.0, so $p = 3000$. In this way, we can calculate the information space radius of a single social individual under Web2.0 conditions as

$$r = \sqrt{\frac{12000}{38.2\pi}} \approx 10(m) \tag{4-26}$$

Then we consider that these social individuals are in a region with a population of 100 million and a land area of 200,000 square kilometers. The entire society is in a space with a height of 3 meters. In addition, considering that social individuals are distributed on the surface of the earth, the information space is also a space with limited height. This height is calculated according to the average height of social individuals as $h_p = 1.7m$.

First, we can calculate the physical space size of this society as:

$$V_p = hS = 3 \times 2 \times 10^{11} = 6 \times 10^{11} \left(m^3\right) \tag{4-27}$$

Then the size of the information space after using Web1.0 is:

$$V_H = h_p n\pi r^2 = 1.7n\pi r^2 = 1.7 \times 10^8 \times \pi \times 3.2^2 \approx 5.5 \times 10^9 \left(m^3\right) \tag{4-28}$$

The size of the information space after using Web2.0 is:

$$V_H = 1.7n\pi r^2 = 1.7 \times 10^8 \times \pi \times 10^2 \approx 5.3 \times 10^{10} \left(m^3\right) \tag{4-29}$$

Of course, for the world's megacities with higher population density, such as Tokyo, its population reaches 37 million, and the entire Tokyo area is only about 2,155 square kilometers. It can be calculated that its physical space is only

$$V_p \approx 3 \times 2155 \times 10^6 \approx 6.5 \times 10^9 \left(m^3\right) \tag{4-30}$$

The size of the information space after using Web2.0 can reach

$$V_H = 1.7n\pi r^2 = 1.7 \times 3.7 \times 10^7 \times \pi \times 10^2 \approx 2.0 \times 10^{10} \left(m^3\right) \quad (4\text{-}31)$$

Therefore, in cities with higher population density, after using advanced information technology, the information space will even exceed the physical space. This is basically consistent with the qualitative analysis of the actual situation. That is, individual information technology in densely populated cities has a greater impact on daily life (Carlson & Goss, 2016). This is called the digital divide (Van Dijk, 2006). This difference is all-encompassing. For example, a study shows that compared with urban doctors, rural doctors use information technology significantly less than urban doctors (Menachemi, Langley, & Brooks, 2007). This digital divide also exists in the education of urban and rural children (Li & Ranieri, 2013). In terms of the frequency of e-commerce use between urban and rural areas, rural areas are also significantly less than urban areas (Zhu & Chen, 2013). Such digital divides exist even within low-productivity countries like Africa (Furuholt & Kristiansen, 2007). These qualitative studies actually show that the densely populated urban information technology provides a larger information space for social individuals, so that urban citizens can make full use of information technology to support their daily life and work.

Although the above data is a very rough estimate and is highly subjective, we can still obtain specific quantitative values of the size of the information space that can only be analyzed qualitatively in the past. To calculate the size of the information space more accurately, it can be determined through more specific data collection under conditions in the future.

4.4 Pressure on biological systems

4.4.1 Composition of social pressure

In physics, pressure is defined as

$$P = \frac{F}{S} \quad (4\text{-}32)$$

For a cube with base area S and length l, then

$$V = Sl \tag{4-33}$$

We can also calculate how much work the pressure does:

$$W = Fl = PV \tag{4-34}$$

Like social space, social pressure can also be divided into physical pressure and information pressure. Physical pressure is an external force exerted on the entire society, causing basic society members or individuals to move. This external force can resist the internal pressure of the social system, or change the size of the social space. Information pressure is an expectation of possible changes within the social system through the way of externally transmitting information. The stronger the expectation of such change, the greater the information pressure. If the social system is changed as a result, information pressure has done its work.

Here, P_p represents physical pressure, and P_H represents information pressure. Then the total pressure on a society is:

$$P = P_p + P_H \tag{4-35}$$

Math further

Combining the physical volume and information volume in social space, we can find that there are four kinds of work, which are

1. Measurable physical work

$$W_p = P_p V_p \tag{4-36}$$

2. Information "work"

$$W_H = P_H V_H \tag{4-37}$$

3. Cross "work"

$$W_{pH} = P_p V_H \tag{4-38}$$

$$W_{Hp} = P_H V_p \tag{4-39}$$

Physical pressure, such as a tsunami, causes social living spaces to be compressed. The force of the tsunami multiplied by the compressed space of the society can obtain the energy released by the tsunami to the society. In other aspects, for example, due to the increase in population, social space tends to increase, but if the land area does not increase, it will lead to an increase in social temperature, which will lead to an increase in physical pressure and social space expansion needs.

Information pressure is an expectation. For example, society expects disasters to occur, which will lead to social migration. This kind of social migration is like the result of the entire social space being compressed after the tsunami. So information can also have a force effect.

Information pressure mainly comes from the interaction between social individuals. We have pointed out in the previous chapters that there are two kinds of interactions among basic society members, namely gender interaction and social interaction. Since in a society, gender molecular or nucleus is a combination of two different gender individuals, the whole society is neutral. Unless there is a serious gender imbalance in a society, otherwise, gender interaction does not cause information pressure changes in the entire social space. Therefore, information pressure mainly comes from social interaction. For example, the accumulation of a large amount of information leads to the emergence of a huge social interaction in a society, and even the formation of a social black hole. Then the society has a huge attraction to the surrounding society.

The definition of physical pressure is very clear and easy to calculate. However, there are certain difficulties in how to define information pressure. This is because the information is relatively abstract and does not meet the requirements of energy conservation.

However, physical and information pressure all can be defined by Newton's laws. If an object gains acceleration, then it is experiencing a force. The magnitude of the force is obtained by multiplying the mass by the acceleration. Of course, acceleration is more difficult to measure, and it is easy to be confused with the curvature of space-time caused by gravitational fields or information fields. Therefore, in the theory of relativity, momentum is more commonly used to define time differentials.

The movement of individuals in society is caused by the energy consumed internally. Momentum changes occur when individuals move because of information. Differentiating time can be seen as the information force received。

$$F = \frac{dp}{dt} \tag{4-40}$$

where p is momentum.

For example, a society is about to be invaded by a foreign enemy, causing panic in the whole society, resulting in a large migration of social individuals. The velocity of the individual's evacuation caused by this foreign invasion information multiplied by the individual's total mass is the momentum.

Considering that in the social system, a large amount of energy of social individuals will be consumed in the neural network of the brain, so information power can also be obtained from the perspective of energy consumption. The formula for work:

$$W = Fl \tag{4-41}$$

where W is the work done by the force F, and l is the distance moved. So

$$F = \frac{W}{l} \tag{4-42}$$

Of course, external work can also be converted into heat, which is exactly what social thermodynamics needs to deal with.

These two calculation methods are used in different situations. The momentum method is mainly suitable for short-term acceleration. The calculation of energy consumption is suitable for long-term individual flow.

4.4.2 Case study: Information force received by customers on Black Friday

Black Friday is a Christmas sale in some malls in western countries such as the UK and the US (Thomas & Peters, 2011). Due to the very large discounts on commodities, many customers will enter the mall at this time to snap up large-scale purchases. Many customers will stand at the door of the store when the store is not open, and when the store is open, these customers will rush into the mall to buy those heavily discounted items.

When the mall opens for business, they rush into the mall to buy and snap up. The motivation here must be the information that the products are greatly discounted. Therefore, by analyzing the momentum changes of these customers after they rush into the mall, we can understand the information of commodity promotions and the magnitude of the force on these customers.

Suppose the number of customers gathered at the entrance of the mall is n, and the mass of a single customer is m. The average distance from the mall gate to the interior of the mall is d. The average speed of customers entering the mall is v.

In this way, we can calculate the momentum of the customer running to the discounted item as

$$p = nmv \tag{4-43}$$

The time required is:

$$\Delta t = \frac{2d}{v} \tag{4-44}$$

Therefore, we can calculate the force of these customers as:

$$F = \frac{\Delta p}{\Delta t} = \frac{nmv^2}{2d} \tag{4-45}$$

If the customer's acceleration distance is $d = 20m$; the customer's weight $m = 60kg$.

Let the number of customers be $n = 100$; the speed after acceleration is $v = 5\ m/s$.

This way we can calculate:

$$F = nm \cdot \frac{v^2}{2d} = 100 \times 60 \times \frac{5^2}{2 \times 20} = 3750(N) \tag{4-46}$$

This is still a relatively large force, and the average force per person is about 62.5*N*.

However, the above calculation only takes into account the force required to accelerate the customer's speed from 0 to v. It doesn't take into account the friction that customers need to overcome as they move back and forth in the purchasing process. This friction is also caused by product discount information. There is another energy that is not accounted for. It is the energy of the brain neural network that the customer needs to process the information when they choose the product.

Therefore we can also solve for this force in terms of overall energy consumption. The average basal metabolic rate of an adult is 1200 kcal. If there is a relatively large amount of exercise, the calories consumed can reach 1700 kcal. Considering that the physical consumption of customers on Black Friday is relatively large, the calorie consumption of panic buying during this period is set at 100 kcal, and the unit converted into energy is 418,400 J.

Then we calculate according to the 50 meters per row of shelves in the shopping mall, a total of ten rows, customers need to complete an average distance of 500 meters for panic buying, so we can calculate the size of the force received by a customer as:

$$F = \frac{W}{l} = \frac{418400}{500} = 836.8(N) \tag{4-47}$$

It can be seen that the force calculated in this way is much larger than the force generated by the momentum change. The reason for this is that the force calculated using the momentum change simply reflects the magnitude of the resultant force required to move the customer. Therefore, it is more comprehensive to use the method of energy consumption to calculate the force received by the customer.

Considering that the external driving force for these customers to rush into the mall is mainly the information of commodity promotions. Therefore, these forces can be classified into the category of "information force". This is not the same as the situation in which the gathering crowd is carried out by security guards or pushed out of the scene by physical force.

Of course, the essence of information force is produced by physical interaction. The customer needs to move and the neural network needs to be impulsive, all of which need to consume the energy of the body.

4.5 Temperature and entropy of biological systems

4.5.1 The meaning of social temperature and entropy

Like social pressure and social space, social temperature can also be discussed from both "physical" and "informational" aspects. However, to avoid confusion between the term "physical temperature" and the atmospheric temperature in the environment we live in, we use "physiological temperature" here instead of the term "physical temperature" of society.

1. Physiological temperature

Physiological temperature reflects the activity level of social individuals. If the individuals in a society are very active and have greater kinetic energy, then the whole social temperature, its physiological temperature part will be relatively high. Physiological temperature is easily represented by various exercise parameters. Take the mass multiplied by the square of the velocity divided by 2, which can be used to reflect the kinetic energy of individuals in a society. The kinetic energy of social individuals is closely related to the activity ability of social individuals, so it is more accurate to use this physiological temperature to represent the society formed by animals. For example, in the social system composed of wolves, because the speed of wolves is very fast, the physiological temperature of the entire wolf society is relatively high.

Math further

In the gas molecular theory, if the kinetic energy distribution of the gas molecular conforms to the Boltzmann distribution, the relationship between its temperature and the molecular average kinetic energy can be expressed by the formula:

$$E_k = \frac{1}{2}m\overline{v}^2 = \frac{3}{2}k_B T \tag{4-48}$$

where k_B is the Boltzmann constant, m is the gas molecular mass, and $\overline{v}$ is the molecular mean velocity.

However, we consider that in the social system, the social individual is active on a plane, so the dimension of the social individual is usually only 2-dimensional, so the above formula is applied to the social system and can be expressed as:

$$E_k = \frac{1}{2}m\overline{v}^2 = k_B T_p \tag{4-49}$$

Therefore

$$T_p = \frac{m\overline{v}^2}{2k_B} \tag{4-50}$$

It can be seen from this that the more active the individuals in a social system, the higher the physiological temperature of the social system. For example, if the work pace of individuals in cities is more intense, the activity of each individual will be significantly larger than that in sparsely populated rural areas. In this way, the kinetic energy of each individual increases accordingly, resulting in an increase in the physiological temperature of the urban society. However, the rise in physiological temperature in this urban society is different from the rise in atmospheric temperature measured by a thermometer. The reason is that the physiological temperature of human society is a macroscopic phenomenon. The change of atmospheric temperature measured by the thermometer is a microscopic phenomenon belonging to the atmospheric molecular motion. Although it has been shown that the average temperature in some large cities is always about three degrees warmer than the surrounding suburbs which are not so densely populated. This higher temperature of 3°C is more likely to be related to the body temperature aggregation effect caused by the aggregation of social individuals. That is, as long as more social individuals come together, even those individuals who are not exercising will also increase the temperature.

2. Information temperature

Different from the social system composed of animal groups, in the human social system, the most important thing is the transmission of information. The transmission of this information often leads to changes in the activity level of the entire society.

However, the size of the amount of information is independent of the activity level of individuals, or the kinetic energy of social individuals. The greater the amount of information that a social individual possesses, does not mean that a social individual has the ability to act, or that the kinetic energy of the social individual will be greater. On the contrary, in our modern society, some individuals who are addicted to games, although their amount of information is very large, they are in a state of very little activity.

Therefore, information transmission itself will also cause changes in social temperature. Due to the different levels of complexity of information, in some relatively primitive societies, the information conveyed is relatively simple. Other societies communicate messages that are highly complex. Different levels of information complexity also reflect the different levels of civilization of a society. For some societies with a very high degree of civilization and very advanced science and technology, the total amount of information transmission is also significantly higher than those of primitive tribal societies that basically have no information transmission.

In order to study the influence of information on social temperature, we can understand information from a more microscopic level. We can then think of the entire society as one giant neural network. This neural network is an internet of neural networks because it connects the neural networks of the brains of all individuals in a society.

The impulse of the neural network of the brain must be accompanied by the energy consumption of the brain. This brain energy consumption is directly related to the activity of neural network impulses. That is to say, the stronger the impulse of the neural network, the higher the energy consumption of the brain. The impulse of the neural network is closely related to the amount of information. If the amount of information to be processed is very large, then the neural network of the brain will generate larger-scale neural network impulses.

If we compare the energy consumption of the brain neural network with the kinetic energy of the gas molecular, we can find that, in fact, the energy consumption of the brain neural network is consistent with the kinetic energy of the gas molecular. Since the kinetic energy of the gas molecular directly reflects the temperature of a physical system, we can also regard the average neural network energy consumption of each social individual in a social system as the average kinetic energy of the individual in this social system. With the average individual kinetic energy of the social system, and this is also a macro system, which conforms to the Boltzmann distribution, we can directly link the temperature of this social system with the energy consumption of the social individual's brain neural network. We call this phenomenon of social temperature caused by the energy consumption of the brain neural network as information temperature. This can be compared with physiological temperature.

For social systems, whether it is social physiological temperature or information temperature, we are temporarily unable to use thermometers to measure. Although studies have shown that there is still a certain correlation between ambient temperature and psychological temperature, the social temperature we are talking about here is more of a macroscopic phenomenon. The temperature measured by the thermometer is caused by the kinetic energy of the microscopic molecular motion.

According to the calculation of thermodynamics, the relationship between internal energy and entropy meets the requirements of formula (2–15) under the condition of constant temperature. However, we have introduced the concept of social temperature here, so in order to avoid confusion, we use α to represent the temperature of physical thermodynamics. So formula (2–15) becomes:

$$dU = \alpha dS = \alpha dH \tag{4-51}$$

Considering that the body temperature of humans and mammals is constant, so

$$\alpha = Const. \tag{4-52}$$

If we regard the energy consumption of the brain as the same parameter as the kinetic energy of the gas, and the impulse of the neural network of the social individual brain has only one dimension, then:

$$E_k = \alpha H = \frac{1}{2} k_B T \tag{4-53}$$

Therefore

$$T_H = \frac{2\alpha H}{k_B} \tag{4-54}$$

This shows that the energy consumption of the brain is consistent with the size of the amount of information. That is to say, the greater the amount of information processed by the brain, the greater the energy consumed by the brain. That is, the change in energy caused by the change in the amount of information is consistent. Therefore, considering the characteristics of Boltzmann distribution, the greater the amount of information transmitted in human society, the higher its information temperature will be.

Of course, the calculation of the amount of information is very complicated. So Equation (4–51) actually also provides a way to estimate information temperature directly using energy consumption.

3. Social entropy

We can also divide social entropy into physical entropy and information entropy. For physical entropy, human society has a very strong ability to integrate, that is to say, when different cultures or different ethnic groups gather together, there may be a kind of social integration. For example, some of these individuals are very active, but if they enter another society, after the passage of time, these individuals will gradually integrate into the whole society. Physically, this actually reflects an increase in the number of states, similar to mixing two substances of different densities. The result of this mixing ultimately leads to an increase in physical entropy. This is in accordance with the first law of thermodynamics. But we should also see that because human society is a system far from equilibrium, and there is constant energy input to maintain the stability of the entire non-equilibrium state. Therefore, this increase in entropy does not mean that the whole society has entered a state that cannot be changed, and the whole society is still full of very powerful vitality.

The relationship between information entropy and amount of information is relatively straightforward. The increase of the amount of information actually means the increase of information entropy. The use of information technology leads to an increase in the total amount of information and an increase in information entropy. At the same time, with the development of human civilization, more and more knowledge and information will be accumulated, which will also lead to an increase in the amount of information that social individuals need to process. This is also a direct cause of the rise in the temperature of information in a society.

4.5.2 Case study: The relationship between climate change and the development of social civilization

Global climate change, especially rising temperatures, is causing a society to store more heat in its physical environment. If social individuals can use these physical energy sources more efficiently, the energy supply of the brain's neural network can be improved. The improvement of the energy supply of the brain neural network of social individuals can improve the ability of the brain neural network to process information. Thereby increasing the temperature of information in society.

Of course, the improvement of the brain's information processing ability also means that the accumulation of knowledge in human society is directly linked to the degree of civilization of a society. From this, we can draw a conclusion that the rise in global temperature has a positive effect on the development of civilization achievements of the entire human society. These civilization achievements include science and technology, literature and art and other aspects.

Here we analyze the impact of temperature changes in Chinese history on the development of Chinese civilization based on the historical development of China in five thousand years. At present, there is still a lot of research on the relationship between ancient China's climate changes and the replacement of dynasties (Fan, 2010). This is mainly due to the fact that compared with other countries, China has relatively complete historical records that can provide evidence of temperature changes over the years.

By comparing the change of temperature with the change of dynasties, some scholars have found (Hinsch, 1988) that when the temperature in China rises, it is also a period of prosperity of dynasties. When the average temperature of the whole country drops, it is generally the period when disasters are rampant, peasant uprisings, the movement of ethnic minorities to the south, and even the change of dynasties.

Among them, Zhu Kezhen is the most in-depth researcher on climate change in ancient China. In 1972 and 1973, he obtained the temperature change curve of ancient China through the analysis of existing archaeological discoveries, literature records and other materials. According to Zhu Kezhen's research results (Zhu, 1973), we can find that the rise and fall of each dynasty is indeed closely related to the average temperature at that time.

Figure 4-1 shows a rough result of this:

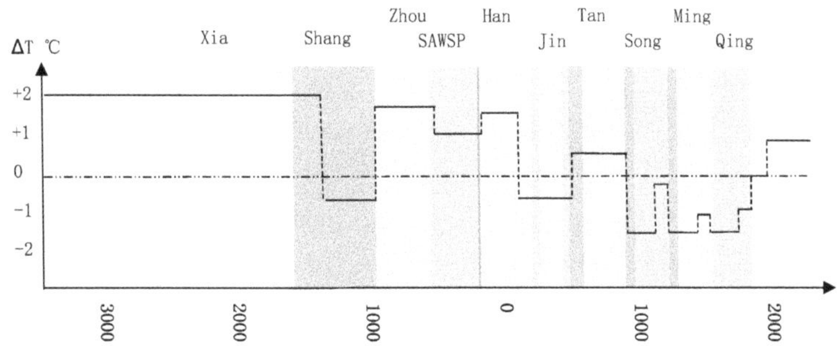

Figure 4.1. Temperature change and China's historical development
SAWSP: Spring and Autumn and Warring States Period

In the figure, there is also the Qin Dynasty between the SAWSP and the Han Dynasty. There are also three kingdoms between the Han and Jin dynasties. Between the Jin Dynasty and the Tang Dynasty, there were the Southern and Northern Dynasties and the Sui Dynasty. There are "Five Dynasties" between the Tang and Song dynasties, and there is the Yuan Dynasty between the Song and Ming dynasties.

As can be seen from the figure, according to climate change, China's 5,000-years history can be roughly divided into two parts. These two parts use the Han Dynasty as a dividing line. Before the Han Dynasty, it belonged to a high temperature stage, and after the Han Dynasty, the temperature of the whole China dropped more seriously. It can be clearly seen from the picture that before the Han Dynasty, it mainly experienced the Xia Dynasty, the Shang Dynasty, the Zhou Dynasty, the Spring and Autumn Period and the Warring States Period, the Qin Dynasty and so on. Except for the Qin Dynasty, the history of these dynasties lasted for a long time, ranging from hundreds of years to thousands of years. This also shows that during this period, the whole society is in a relatively stable stage. During this period, many other countries in the world were still in the very primitive Stone Age, but China's civilization achievements in this period of history were very advanced. The transition from the Neolithic to the Bronze Age to the Iron Age was successfully completed. This leap is very huge, and it also marks the beginning of the budding of the achievements of modern civilization. But after the Han Dynasty, we can notice that the temperature throughout China began to drop, and this drop in temperature also brought about greater fluctuations in temperature. The large fluctuations in temperature actually mean that the energy supplied to the entire society fluctuates wildly. This also led to the frequent changes of dynasties in Chinese society during the 2,000 years from the Han Dynasty to the modern era. These dynasties were as short as a few decades and as long as three or four hundred years. Compared with the Shang Dynasty and other ancient Chinese dynasties in the past, it is far inferior. From this, we can draw a conclusion that when the temperature in the natural environment rises, it means that the social individuals can be given more sufficient energy. The provision of this sufficient energy will make the brain neural network of social individuals more active, and will cause the temperature of information to rise. This rise in information temperature is the basis for the development of various civilization achievements.

Judging from the development of science and technology in ancient China, before the Han Dynasty, the development of science and technology in ancient China was actually very fast. Completed the invention of writing and the improvement of metal smelting technology. Especially in the Spring and Autumn Period and the Warring States Period, the development of this kind of science and technology reached its peak. Philosophers, educators and scientists such as Laozi, Confucius, Mozi and so on have emerged. But after the Han Dynasty, the development of Chinese science and technology basically stagnated in the Iron Age. After the Han Dynasty, although a group of philosophers and artists also emerged, it is difficult to continue to break through the peak of science and technology in the Spring and Autumn Period and the Warring States Period in the development of science and technology. As a result, after the 19th century, China's scientific and technological development was significantly behind that of Western countries such as Europe and the United States.

Although the development of science and technology in China was very slow after the Han Dynasty, and there were not many innovative technologies and theories, we can also notice that there were some fluctuations in temperature during this period of relatively low temperature. When a dynasty is relatively prosperous, all kinds of literature, art, technology, and even scientific theory have developed to a certain extent at this time. This development corresponds to the rise in information temperature. This information temperature rise has a great positive relationship with the overall temperature rise at that time. For example, we can see that during the heyday of the Tang Dynasty, the average temperature in China was still relatively high. But when the temperature began to drop, at this time we saw that the Tang Dynasty began to decline, and the dynasties began to change, changing to dynasties such as the "Five Dynasties" and the Song Dynasty. At the beginning of the Song Dynasty, the temperature was relatively low, but at the end of the Northern Song Dynasty, the temperature of the whole China began to rise again. This means that information temperature is also starting to rise. We can also notice a similar situation in other dynasties. For example, although the temperature in the Ming and Qing dynasties has reached the lowest point, there are still some periods when the temperature will rise in these two dynasties. And this kind of temperature increase, we can find that literature, art, science and technology still have a certain degree of development during this period.

Human activities in ancient societies are less likely to change atmospheric temperature. After all, there were basically no industrial technologies at that time, and there were relatively few sources of these greenhouse gases that produce carbon dioxide or methane (Smith et al., 2008). But after the 20th century, the situation changed, and human activities began to change the climate. When human activities are more intense, especially the rapid development of industry, a large amount of greenhouse gases are discharged into the atmosphere. These greenhouse gases, including carbon dioxide, methane, etc. (Lashof & Ahuja, 1990), caused the global temperature to begin to rise rapidly. Atmospheric temperature has risen by more than 1°C on average over the past 100 years (Masson-Delmotte et al., 2018). So that mankind has to adopt a global cooperative approach to limit greenhouse gas emissions (Burns, 1997). This rise in temperature is due to human activity. Of course, it can also be explained from another point that when human civilization develops to a certain extent, the continuous increase of information temperature will lead to the rapid development of industry, and the rapid development of this industry will directly affect the temperature increase. As a result, social temperature is more closely related to atmospheric temperature, and the rising and falling trends between the two are more consistent.

References

Burns, W. C. (1997). Global warming-the United Nations Framework Convention on Climate Change and the future of small island states. *Dickinson Journal of Environmental Law & Policy*, 6, 147.

Carlson, E., & Goss, J. (2016). The State of the Urban/Rural Digital Divide. https://www.ntia.doc.gov/blog/2016/state-urbanrural-digital-divide. August 10.

Fan, K. W. (2010). Climatic change and dynastic cycles in Chinese history: A review essay. *Climatic Change*, 101(3), 565–573.

Furuholt, B., & Kristiansen, S. (2007). A rural–urban digital divide? Regional aspects of Internet use in Tanzania. *The Electronic Journal of Information Systems in Developing Countries*, 31(1), 1–15.

Heuser, S. (2003). Neal Stephenson's metaspace. In *Virtual geographies* (pp. 171–189). Brill Rodopi.

Hinsch, B. (1988) Climatic change and history in China. *Journal of Asian History* 22(2), 131–159.

Lashof, D. A., & Ahuja, D. R. (1990). Relative contributions of greenhouse gas emissions to global warming. *Nature*, 344(6266), 529–531.

Li, Y., & Ranieri, M. (2013). Educational and social correlates of the digital divide for rural and urban children: A study on primary school students in a provincial city of China. *Computers & Education*, 60(1), 197–209.

Masson-Delmotte, V., Zhai, P., Pörtner, H. O., Roberts, D., Skea, J., Shukla, P. R., ... & Waterfield, T. (2018). Global warming of 1.5 C. An IPCC Special Report on the impacts of global warming of 1.5°C, 1(5).

McPherson, S. S. (2009). *Tim Berners-Lee: Inventor of the World Wide Web.* Twenty-First Century Books.

Menachemi, N., Langley, A., & Brooks, R. G. (2007). The use of information technologies among rural and urban physicians in Florida. *Journal of Medical Systems*, 31(6), 483–488.

O'reilly, T. (2009). *What is Web2.0.* O'Reilly Media, Inc.

Smith, P., Martino, D., Cai, Z., Gwary, D., Janzen, H., Kumar, P., ... & Smith, J. (2008). Greenhouse gas mitigation in agriculture. *Philosophical Transactions of the Royal Society B: Biological Sciences*, 363(1492), 789–813.

Thomas, J. B., & Peters, C. (2011). An exploratory investigation of Black Friday consumption rituals. *International Journal of Retail & Distribution Management*, 39(7), 522–537.

Van Dijk, J. A. (2006). Digital divide research, achievements and shortcomings. *Poetics*, 34(4–5), 221–235.

Zhu, K. (1973) A preliminary study on the climatic fluctuations during the last 5000 years in China. *Scientia Sinica*, 16, 226–256

Zhu, S., & Chen, J. (2013). The digital divide in individual e-commerce utilization in China: Results from a national survey. *Information Development*, 29(1), 69–80.

5

Basic social thermodynamic equations

5.1 Overview

5.1.1 Fundamental laws of thermodynamic systems

With the definition of important parameters such as social pressure, social space and social temperature, we can use thermodynamic methods to solve social problems. In social thermodynamics, the transmission of energy and information is mainly involved, so solving the problem of social thermodynamics also mainly involves how to calculate the transmission of information and energy and the work done by various forces. The treatment of social thermodynamics is basically the same as that of thermodynamics, except for the parameter involved in information.

Since these parameters introduced in social thermodynamics have a one-to-one correspondence with thermodynamics, then we can believe that the laws of nature that can be applied in thermodynamics can also be applied to social thermodynamics. And with the exact same units as those in thermodynamics, when dealing with social thermodynamics, we can also directly apply some important laws and formulas in thermodynamics. These important laws in thermodynamics mainly include four laws of thermodynamics. These laws of thermodynamics are universally applicable in thermodynamics.

There are four most basic laws of thermodynamics, corresponding to the 0th law of thermodynamics to the 3rd law of thermodynamics.

The 0th law of thermodynamics means that for two contacting thermodynamic systems, if their temperatures are different, then heat will be transferred from the thermodynamic system with the higher temperature to the thermodynamic system with the lower temperature. Eventually the temperature of the two thermodynamic systems is equalized.

If other factors remain constant, the relationship between the heat transferred and the temperature difference is

$$dQ = SdT \tag{5-1}$$

The first law of thermodynamics is essentially the law of conservation of energy. It reflects how mechanical energy is converted into thermal energy in a thermodynamic system.

That is

$$dU = dW + dQ \tag{5-2}$$

The second law of thermodynamics reflects the change law of the entropy of a thermodynamic system. The second law of thermodynamics states that in an equilibrium thermodynamic system, entropy is always changing in the direction of maximization.

That is

$$\delta S \geq 0 \tag{5-3}$$

The third law of thermodynamics states that for any thermodynamic system, it is impossible to reach $0K$.

$$T > 0K \tag{5-4}$$

5.1.2 The applicability of the laws of thermodynamics in social thermodynamic systems

Because the important parameters such as social pressure, social temperature and social space have been accurately defined in Chapter 4, theoretically, all the laws used in thermodynamics can be directly transplanted into the social thermodynamic system. However, when applying it, we can still pay attention to some problems that need to be paid attention to when applying the laws of thermodynamics in the social thermodynamic system.

For the basic equations of thermodynamics, especially the basic equations of ideal gases, some of the constants involved may be different from the thermodynamic system. For example, Boltzmann constant, Avogadro constant and so on. These constants are mainly calculated for the microscopic state of the matter system. But for the social system, the social individuals involved are much more complex than the atoms and molecules in the matter. In addition, the number of social individuals is significantly less than the number of atoms and molecules in the matter system. Even for a country with a very large population like India or China, the total population is only a few billion. But for gas molecules, any milliliter of gas can contain up to 10 to the 20th power. Therefore, in the matter system, we can use the knowledge of statistical physics or mathematical statistics to accurately predict the future trend of the system, and obtain the macroscopic parameters of these microscopic systems from these statistical average values. Considering the limited number of social individuals, especially for some relatively small societies, there may be only a few thousand or even hundreds of individuals, so the statistical effect of such a small number is relatively weak. This is not to say, of course, that statistical physics methods cannot be used in social systems. We only need to consider that when using these statistical physical methods, there will be larger errors, and we can also obtain the macro parameters of these social systems. Compared with the thermodynamic system and the social thermodynamic system, the larger error of the macroscopic parameters can also bring some very important characteristics different from the physical thermodynamic system. These are the important factors that need to consider when testing the correct application of thermodynamic methods in social thermodynamic systems.

Specifically, for the ideal gas equation of state in a thermodynamic system, since the number of individuals in the thermodynamic system is relatively small, the error of the macroscopic parameters will be relatively large. In this way, we can notice the influence of the abnormal situation of some social individuals on the calculation results when the equation is applied. At the same time, in the social thermodynamic system, the interaction between social individuals is often much larger than the interaction between atoms or molecular, so when dealing with specific problems, the social thermodynamic system may be more suitable for liquid or solid models.

For the zeroth law of thermodynamics, this law reflects that the temperature between two systems in contact will gradually tend to be the same. Considering that the number of individuals in the social system is relatively small and the error of the macroscopic parameters is relatively large, the heat transfer efficiency between the two social thermodynamic systems may not be as high as that of the physical thermodynamic system. In this case, the heat transfer in the social thermodynamic system may last for decades or even centuries.

For the first law of thermodynamics, this law involves the conservation of energy. Chief among them is the conversion between mechanical work and heat. In the social thermodynamic system, the energy conversion involved is far more complicated than the matter thermodynamic system. This involves a parameter such as information, so information pressure, information space and information temperature are derived. Information is not the same as energy. Information often does not obey the law of conservation. This means that information can be copied indefinitely. In this way, when it comes to the first law of social thermodynamics, it is necessary to take into account the energy consumption of these information pressures and changes in information space. This conversion of energy is indirect compared to the thermodynamic system of matter.

For the second law of thermodynamics, it involves a change in entropy. For a thermodynamic system of matter, if it is in equilibrium, this means that the entropy of the system will eventually reach its maximum value over time. A social thermodynamic system is different from a physical thermodynamic system. Each individual in a social thermodynamic system is actually a complex system far from equilibrium. Entropy will be kept extremely small at all times while relying on continuous energy input. Therefore, when applying the second law of thermodynamics to the social thermodynamic system, we are actually standing on a higher macro level to consider the changes of various thermodynamic parameters of the entire social system.

For the third law of thermodynamics, it is impossible for any system of matter to reach absolute zero temperature. This is more evident in the social thermodynamic system. Because in this social system, unless all social members have disappeared, the system will always be far from the equilibrium state, and maintain sufficient interaction between individual members of the society. And this sufficient interaction is a basis for maintaining a sufficient social temperature of the entire social system.

5.2 Equation of state

5.2.1 The PVT equation for social systems

The social thermodynamic system is more complex than the physical thermodynamic system. In addition to the more diverse interactions, what is faced in social thermodynamics are macro-individuals. Each individual contains a very large number of molecules. A social individual is even larger than the system studied in physical thermodynamics. Therefore, if the molecular number contained in the whole social individual is directly used, it will bring great complexity to the calculation process.

From the derivation process in the previous section, we can see that although thermodynamic itself deals with macro parameters, it still involves some very important micro parameters. Such as Avogadro constant, Boltzmann constant and so on. Using these microscopic constants means that in a social thermodynamic system we also have to deal directly with molecular and atomic.

To obtain the macroscopic equation of state of social thermodynamics, the main thing is to transform the Avogadro constant or Boltzmann constant.

Therefore, we can slightly improve the various formulas in the equation of state, and the ideal gas equation of state becomes the following form:

$$pV = NnRT \tag{5-5}$$

In this new formula, N represents the number of individuals in the society, and n represents the number of moles contained in a single social individual. If n is the number of moles, the constant is R

Of course, we can also take n as the molecular number contained in a single social individual, then the constant is the Boltzmann constant, and the formula becomes:

$$pV = Nn_p k_B T \tag{5-6}$$

where N is the number of social individuals, n_p is the molecular number contained in a single social individual, and k_B is the Boltzmann constant.

We go one step further and change the equation of state for an ideal gas to:

$$pV = N\beta T \tag{5-7}$$

where β is a new constant. The constant is related to the type of organism, then

$$\beta = n_p k_B \tag{5-8}$$

For an actual social thermodynamic system, we can also change the Van der Waals equation to

$$\left(p+\frac{a_s N^2}{V^2}\right)(V-b)=N\beta T \qquad (5\text{-}9)$$

In the above equation, if the social space occupied by social individuals and external pressure are not considered, it can be changed to

$$\frac{a_s N^2}{V^2}V=N\beta T \qquad (5\text{-}10)$$

After simplification we get

$$a_s N=\beta TV \qquad (5\text{-}11)$$

It can be seen that under the condition that the social space is constant, if the Van der Waals constant a_s is constant, the social temperature is proportional to the number of social individuals.

And if N does not change, the social temperature is proportional to a_s. Considering that the constant as reflects the strength of social interaction, it also means that in the absence of external pressure, the stronger the social interaction, the higher the social temperature. If you want to lower the social temperature, you can achieve the goal by expanding the social space.

5.2.2 Estimation of parameters

The human body itself is a thermodynamic system, and the main content of the human body is water. Therefore, it is assumed that the entire human body is composed of water molecules, so that the number of molecules contained in a human body can be easily estimated.

Considering that the volume of a person is about 0.06 m^3, and the number of molecules contained in every 18 grams of water is 6.02×10^{23}, so the number of moles contained in 1 cubic meter of water is about

$$n_{water}=\frac{10^6}{18}\times6.02\times10^{23}\approx3.3\times10^{28}$$

From this, the parameter n_p can be estimated, that is, the molecular number of waters contained in a human body is approximately:

$$n_p = 0.06 \times 3.3 \times 10^{28} \approx 2 \times 10^{27} \tag{5-12}$$

According to this calculation, we can also estimate the constant of human society

$$\beta = 2 \times 10^{27} \times 1.38 \times 10^{-23} = 2.76 \times 10^{4} \left(J \cdot K^{-1}\right) \tag{5-13}$$

Math further

In the Van der Waals equation, there are also two constants "a" and "b". For different types of society, the interaction between social individuals within it is different, which will affect the interaction between social individuals, and then affect social pressure. This social pressure can be determined by calculating the Van der Waals constant a.

The Van der Waals constant "a" reflects the interaction strength between the molecules. This is a constant with a very complex unit. Its unit is " $N \cdot m^4$ *" or "* $Pa \cdot m^6$ *", which can also be expressed as:*

$$Pa \cdot \frac{L^2}{mol^2}$$

Since

$$1Pa = 1\frac{N}{m^2}$$

$$1kPa = 1000Pa$$

$$1L = 10^{-3}m^3$$

$$1mol = 6.02 \times 10^{23}$$

In thermodynamics, the parameter "a" is different for different types of gases. This parameter is usually measured experimentally. For example, the "a" value of H_2O *is:*

$$a = 553.7kPa \cdot \frac{L^2}{mol^2} \tag{5-14}$$

Some noble gases have much lower a values. For example, helium has only:

$$a = 3.46kPa \cdot \frac{L^2}{mol^2} \tag{5-15}$$

In order to facilitate the calculation and comparison, we can convert the above units to get

$$a = 553.7kPa \cdot \frac{L^2}{mol^2} = \frac{5.537 \times 10^{5-6}}{6.02^2 \times 10^{46}} Pa \cdot m^6$$

So

$$a \approx 1.53 \times 10^{-48} Pa \cdot m^6 \tag{5-16}$$

Of course this is a parameter involving microscopic constants. To use in a social thermodynamic system, we can also do some conversions. Considering the pressure in Van der Waals caused by the interaction between the molecules is:

$$\frac{a(Nn_p)^2}{V^2} = \frac{(an_p^2)N^2}{V^2} \tag{5-17}$$

And n_p is the average molecular number possessed by a single human body. Therefore, it is only necessary to multiply n_p^2 to obtain the macroscopic Van der Waals constant in the social thermodynamic system, namely

$$a' = \frac{553.7 \times 4 \times 10^{54}}{6.02^2 \times 10^{46}} \approx 6.1 \times 10^9 kPa \cdot L^2$$

Which is

$$a' = 6.1 \times 10^6 Pa \cdot m^6 \tag{5-18}$$

Since the Van der Waals constant multiplied by the square of the density of individuals in society is pressure. This also shows that the Van der Waals constant a' actually reflects a proportion of the contribution of social individual density to internal pressure. If the constant a' is larger, it means that the density of social individuals is larger, which will generate greater internal pressure. This actually means that the interaction between social individuals is greater. This is consistent with the meaning of social interaction between individuals and gender interaction coupling coefficient. Therefore, the macroscopic Van der Waals constant a_s can also be regarded as the interaction coupling coefficient.

But the intensity of interaction between individuals in human society is obviously much larger than that between water molecules. Intuitively, the water droplets are easy to separate. But if it is a human society, each individual may interact with other people in the social system and form a relatively strong relationship. Such as employment, if a person already has a stable job, it takes a very large interaction force to let this person leave his job.

We do a simple comparison here. Assuming that in the weightless environment of the space station, two water droplets with the same volume as the human body are fused together, then according to the Van der Waals constant a, the internal pressure of the fused water droplets can be calculated as:

$$p_i = \frac{aN^2}{V^2} = \frac{1.53\times10^{-48}\times16\times10^{54}}{4\times0.06^2} = 1.7\times10^{9}\,(Pa) \tag{5-19}$$

This is a very large force. But this is the force needed to break up all the molecular in the water. In fact, we only need to split the droplet in half, so the number of molecules that need to be split is much less.

Considering that the diameter of each water molecular is $4\times10^{-10}\,m$, the sum of the two water molecular diameters is $8\times10^{-10}\,m$.

A spherical water drop with a volume of 0.12 cubic meters and a central cross-sectional area of 0.29 square meters, if the thickness of the cross-section is $8\times10^{-10}\,m$, the volume of this part is

$$V_s = 0.29\times8\times10^{-10} = 2.32\times10^{-10}\left(m^3\right) \tag{5-20}$$

Then we have

$$\frac{V_s}{V} = 1.9 \times 10^{-9} \tag{5-21}$$

This is actually the ratio of the number of moles that need to be separated on the section to the total number of all water moles.

Therefore, the pressure is only about

$$p = 1.9 \times 10^{-9} p_i = 1.9 \times 10^{-9} \times 1.7 \times 10^{9} \approx 3.3(Pa) \tag{5-22}$$

It can be seen that if the two drops of water are coupled together, the pressure required is very small. Compared with the results we calculated in Section 4.4.2, it shows that the force of information on a person can reach 836.8*N*. The pressure of the water droplets glued together is indeed much smaller. Here, if the area of the human body is calculated according to 1 square meter, the information pressure acting on a person can reach 836.8*Pa*. That is to say, in the human social system, the information pressure between people is about 250 times that of the Van der Waals pressure of water.

In this way, we expand the macroscopic Van der Waals constant a_s in formula (5–18) by another 250 times as the coupling coefficient of social interaction in human society, that is, the Van der Waals constant in human social system is:

$$a_s = 250a' \approx 1.5 \times 10^{9}\, Pa \cdot m^{6} \tag{5-23}$$

In addition, considering that gender interaction is stronger than social interaction coupling coefficient, maybe we can expand the social coupling coefficient by 1000 times, and then we can get the Van der Waals constant of gender interaction coupling coefficient as:

$$a_g = 1000a_s \approx 1.5 \times 10^{12}\, Pa \cdot m^{6} \tag{5-24}$$

Of course, in different societies, we will find that the coupling coefficient of this interaction will vary greatly. But we are also facing a problem now that such parameters are actually theoretical estimates, so compared with the actual parameters, the error may be very large. Here we can use a method widely used in physics, which is the method of order of magnitude estimation. That is, if one quantity is larger than the other, physicists are always used to estimating by how many orders of magnitude larger. This can reduce the distraction of thinking about problems caused by a large number of precise calculations.

Therefore, in the subsequent content, for the estimation of some different types of interaction coupling coefficients, we will estimate according to how many orders of magnitude are larger or smaller. For example, the social interaction coupling coefficient of a certain social type is larger than that of another social type. Then we can multiply or divide the corresponding coupling coefficient by 10 times, 100 times, or even 1000 times.

5.2.3 A few examples

Let's do some simple calculations through a few examples to see how the equation of state of the social thermodynamic system is applied. Of course, these examples are all hypothetical, and there is not necessarily a correspondence between the actual history and the actual social situation.

Example 5-1.

Suppose there is a city with a population of 1 million, the external pressure is 10,000Pa, the social temperature is maintained at 272K, try to calculate the size of the city's social space.

Solution:

Since the social pressure outside the city is 10,000Pa and the social temperature is 272K, substituting into ideal gas equation of state (5-7) can get:

$$pV = N\beta T$$

That is

$$10000V = 10^6 \times 2.76 \times 10^4 \times 272$$

Then we get

$$V \approx 7.5 \times 10^8 \left(m^3\right)$$

Calculated according to the height of the social space of 3 meters, that is, 1 million people are actually concentrated within 250 square kilometers.

Example 5-2:

Assuming the city with a population of 1 million in Example 5-1, the shape of the entire city is circular, and the space height is calculated as 3 meters. Assume that all the population of the city is evenly distributed over 1,000 square kilometers. If keep the external pressure constant, try to calculate the change in social temperature.

Solution:
According to Charlie's law, there is a relationship under constant pressure:

$$\frac{V_1}{T_1} = \frac{V_2}{T_2} \tag{5-27}$$

Therefore

$$T_2 = \frac{V_2}{V_1} T_1 = \frac{1000 \times 3 \times 1000^2}{250 \times 3 \times 1000^2} \times 272 = 1088(K) \tag{5-28}$$

It can be seen that if the external pressure remains unchanged and the social space increases, it means that the social temperature will also rise synchronously. Of course, this calculation result shows that this temperature has far exceeded the ambient temperature. So the city has actually had a hard time getting energy freely from the environment. Unless the city is industrialized. Such as the establishment of a large number of factories, and these factories can evacuate the population of the city center to the surrounding suburbs.

Example 5-3:

If the city in Example 5-1 encountered a tsunami by the sea, the population of the city had to move to within 100 square kilometers. Assuming that the social temperature in this migration process is constant, calculate the change in social pressure.

Solution:
From Boyle's law, we can get:

$$p_1V_1 = p_2V_2 \tag{5-28}$$

So social pressure after the tsunami will become

$$p_2 = \frac{p_1V_1}{V_2} \approx \frac{10000 \times 250 \times 1000^2 \times 3}{100 \times 1000^2 \times 3} = 25000(Pa) \tag{5-29}$$

The calculations showed that social pressure increased by the same factor.

Example 5-4:

If you don't want to change the social space of citizens, the pressure of the upcoming tsunami will cause the social temperature to rise. Try to calculate how much the social temperature will rise after the social pressure doubles?

Solution:
According to Guy-Lussac's law, in the case of constant volume, there is a relationship:

$$\frac{P_1}{T_1} = \frac{P_2}{T_2} \quad (5\text{-}30)$$

Then after social pressure rises:

$$T_2 = \frac{P_2}{P_1} T_1 = 2T_1 = 544(K) \quad (5\text{-}31)$$

This rise in social temperature also exceeds the ambient temperature.

Example 5-5.

If the population of a city is 30 million, and the area of the city is 2,000 square kilometers, the external pressure of the city is 0. $a_s \approx 1.5 \times 10^9 Pa \bullet m^6$. Calculate the social temperature of the city. If the city uses Web2.0 Internet technology, calculate the city's social temperature.

Solution:

Since the external pressure is small, the Van der Waals equation is used for the calculation.

$$p_i = \frac{a_s N^2}{V^2} \approx \frac{1.5 \times 10^9 \times 9 \times 10^{14}}{9 \times 4 \times 10^{18}} \approx 3.75 \times 10^4 (Pa) \quad (5\text{-}32)$$

Considering that the social space of each social individual is about 0.06 cubic meters, which is much smaller than the urban area. Therefore

$$V \gg b$$

This can be obtained from the Van der Waals equation:

$$p_i V \approx N\beta T \quad (5\text{-}33)$$

This calculates:

$$T \approx \frac{p_i V}{N\beta} = \frac{3.75 \times 10^4 \times 3 \times 2000 \times 10^6}{3 \times 10^7 \times 2.76 \times 10^4} \approx 272(K) \tag{5-34}$$

This temperature is slightly lower than the ambient temperature. Therefore, even if the city does not use information technology, it can still easily obtain energy from the external environment.

However, after using Web2.0 technology, it can be calculated that its physical space has

$$V_p = 3 \times 2 \times 10^3 \times 10^6 = 6 \times 10^9 \left(m^3\right) \tag{5-35}$$

According to formula (4-29), the information space size after using Web2.0 can reach:

$$V_H = b = 1.7 n \pi r^2 = 1.7 \times 3 \times 10^7 \times \pi \times 10^2 \approx 1.6 \times 10^{10} \left(m^3\right) \tag{5-36}$$

Social space adds information space to physical space. Which is

$$V = 2.2 \times 10^{10} \left(m^3\right) \tag{5-37}$$

so

$$p_i = \frac{a_s N^2}{V^2} = \frac{1.5 \times 10^9 \times 9 \times 10^{14}}{4.8 \times 10^{20}} \approx 2.8 \times 10^3 \,(Pa)$$

It can be seen that the internal social pressure has decreased.

$$T \approx \frac{p_i (V - b)}{N\beta} = \frac{2.8 \times 10^3 \times 6 \times 10^9}{3 \times 10^7 \times 2.76 \times 10^4} \approx 20(K) \tag{5-38}$$

It can be seen that due to the increase of social space, the social temperature can be reduced to a lower level.

This is a very interesting example. For a city, if there is no external pressure, then it depends entirely on internal social interactions, and its social temperature and social space can also be calculated through the Van der Waals equation.

However, from the calculation results, it seems that such a system is not stable. That is, if the social space is continuously compressed due to the existence of internal pressure, the internal pressure will be greater in the process of the social space being compressed, which will make the social space more severely compressed. However, after we made further transformations, we can find that the relationship between temperature and social space is an inverse relationship.

$$T \approx \frac{p_i V}{N\beta} = \frac{a_s N}{\beta V}$$

In other words, if social space is reduced, it will directly lead to an increase in social temperature. This increase in social temperature will in turn lead to a reduction in the possibility of the entire society accessing energy from the environment. Therefore, social individuals in the society will consciously maintain an appropriate social space, so as to prevent the temperature of the whole society from rising to the extent that it is difficult to obtain energy.

5.2.4 Case study: Social temperature of wolves and population size limits

Since, unlike humans, the information exchange between animals is far less complex than humans, there are fewer information issues involved. Some approximations can be made when calculating.

Consider that the communication of information between wolves is far simpler than that of humans. Therefore, the social interaction related to information transmission within the wolf pack is actually very weak. But wolves are mainly organized in families. Therefore, although the social interaction within the wolf pack is relatively weak compared with humans, the gender interaction is very strong. This results in a very strong structure of wolves. If an external force were to break up the wolves, it would take a very large amount of energy to be able to do it. In the matter world, it's like comparing water to very viscous grease. Considering that the social system composed of wolves is relatively close to human society, we use the gender coupling coefficient to calculate the strength of social interaction within the wolf group. That is, take $a_w \approx 1.5 \times 10^{11} Pa \cdot m^6$.

Although there is evidence that a wolf pack can reach a range of 100–300 km (Van Ballenberghe, Erickson, & Byman, 1975). However, in order for the wolves to form a whole through internal gender interaction, it is impossible for the individuals within the wolves to gather in such a large range. The reality is that the reason why a wolf pack is called a "pack" is that the individuals in it are always clustered in a relatively small area. The wider range of activities of wolves refers to the movement range of the entire wolf group. It is like a drop of liquid flowing continuously on a large surface, and the diameter of the liquid bead itself is very small. Therefore, according to such an estimate, the social space occupied by the wolves itself is within the range of about 1 km^2. Calculated according to the space height of 2 meters, the size of the social space of the wolf pack can be obtained as: V=2000000 m^3.

If the social pressure outside the wolf pack is very small, then we can calculate a typical wolf pack with 7 wolves according to formula (5-11), and its social temperature is:

$$T = \frac{a_w N}{\beta V} = \frac{1.5 \times 10^{11} \times 7}{2.76 \times 10^4 \times 2 \times 10^6} \approx 19(K) \tag{5-39}$$

This temperature is very low, which means that wolves with a lower social temperature will more easily obtain energy from the environment. In addition to the typical 7 wolves in the actual wolf pack, there are also some wolf packs with more individuals. If the number of individuals in a pack reaches 28 wolves, its temperature will rise to

$$T = 76\text{K} \tag{5-40}$$

If the number of individuals in the wolf pack continues to rise, the social temperature of the wolf pack will rise further. This also means that it will be much more difficult for wolves to harvest energy from the environment. For example, if the number of individuals in a wolf pack is increased tenfold to 280 wolves, its temperature will reach 760K, which is much higher than the ambient temperature of about 300K. At this point the wolves may release energy into the environment. This situation may exist under some captive conditions.

At the same time, the increase in the number of individuals in the wolf pack may also lead to the weakening of the gender interaction coupling coefficient. These are not conducive to the increase in the number of wolves.

Of course, such calculation results also require that the wolves must be at the top of the food chain, and can freely and unrestrictedly obtain energy in the environment. But in fact, in many cases, wolves are not at the top of the food chain. In particular, human activities, as well as some other problems such as lack of food resources, desertification of the environment, etc., have caused the wolves to obtain energy from the environment, which is greatly limited. These are all causes of increased external pressure on the wolves. From the PVT equation, if the external pressure increases, the living space V will decrease. In the case of adiabatic compression process, the reduction of the living space V means that the social temperature inside the wolf pack increases. This rise in social temperature, according to the zeroth law of thermodynamics, means that the ability of wolves to harvest energy from the environment will decrease. This will result in the number of wolves having to be reduced in order to expand the social space and reduce the temperature of the wolves. For example, there are actually only 2,000 wolves within an area of nearly 100,000 square kilometers on the Hulunbuir grassland.

5.2.5 *Thermodynamic zeroth law and society*

The zeroth law of thermodynamics involves two systems at different temperatures, and energy will be transferred from a system with a higher temperature to a system with a lower temperature. In social thermodynamic, the same units as thermodynamic are used, including social space, social pressure and social temperature, indicating that they have the same meaning. The same dimension also means that the various parameters in social thermodynamic and thermodynamic can call each other. In this way, the results calculated in thermodynamic can be directly used in social thermodynamic. The results obtained in social thermodynamic can in turn be used directly in thermodynamic.

We have calculated social temperature earlier, and its unit is Kelvin. Of course, this social temperature has the same meaning as the physical temperature in thermodynamic. For example, the temperature of the wolf pack we calculated earlier is only 19*K*, which is obviously very low compared to the atmospheric temperature in the environment where animals live in thermodynamics. Human society is the same. A society can maintain such a low social temperature because human or animal society is a system far from equilibrium. In this system, due to the continuous input of energy, human society can maintain a very low social temperature for a long time.

The social temperature of a society composed of animals or people is much lower than the ambient temperature, which is of great benefit. According to the zeroth law of thermodynamics, heat or energy will be transferred from a system with a higher temperature to a system with a lower temperature. This means that the social temperature of human beings is relatively low and the temperature of the environment is relatively high, so that humans can more effectively obtain various energy from the environment to meet and promote the development of human society. Of course, people are different from animals, and people have stronger initiative. Therefore, with the continuous expansion of the scale of human society, the social temperature of human society will continue to rise. Humans can still lower their social temperature in various ways. For example, through various technological advancements, human social space can be effectively expanded. This can lower the overall social temperature. In addition, even if the social temperature of human society is higher than the ambient temperature, it can actively absorb energy from the environment by consuming additional energy through industrial technology. But for animals there is no such ability. Therefore, under the condition that the social space of human beings continues to expand, the number of human beings continues to increase. In this way, humans will occupy the space where many animals live. The adiabatic compression of the living space of animals will also lead to a significant increase in the temperature of the society composed of animals. The great increase in the social temperature of the animal society leads to a rapid decline in the ability of animals to obtain energy from the environment. In order to continuously obtain energy from the environment, the way animals can take is to reduce their number in order to lower their social temperature.

Because all kinds of animals and plants are also an important part of the whole earth's ecological environment. Without a sufficient number of animals, the entire ecological chain of the earth will be severely damaged, which may lead to an increase in the social temperature of human society. Therefore, in our modern society, on the one hand, the rapid development of information technology and industrial technology has improved the survival ability of human beings, on the other hand, we should also pay attention to protecting the society composed of those animals. So that the society of these animals can also freely obtain energy from the environment.

The impact of the zeroth law of thermodynamics on social thermodynamic systems is also reflected in economics. Since the social temperature is lower than the temperature of the environment, this means that the social system is freer to obtain energy from the environment. The composition of these energies is very complex, and for human society, the acquisition of these energies usually exists in the form of commodities.

This means that if the social temperature of the social thermodynamic system is significantly lower than the ambient temperature, people will be more willing to obtain various energy sources from the environment, which is reflected in economics as people's willingness to consume will increase significantly. If the social temperature is relatively high, or even close to the ambient temperature, it is not so convenient for people to obtain energy from the environment at this time, which is reflected in economics and shows that people's willingness to consume is relatively low.

5.3 The first law of thermodynamics

5.3.1 Energy conservation and internal energy

In thermodynamics, energy conservation needs to satisfy

$$dU = dQ + dW \tag{5-41}$$

This means that the work done by the outside world to the system, plus the absorption of heat from the outside world, results in an increase in the internal energy of the system.

And

$$dW = -pdV \tag{5-42}$$

Here the work dW is related to the path.

In thermodynamics, internal energy dU refers to the kinetic energy of the gas molecular. In social thermodynamics, it is considered that temperature is not only related to the physical kinetic energy of social individuals, but also to the size of the amount of information. Therefore, the internal energy in social thermodynamic refers to the sum of the kinetic energy of the social individual and the average energy consumed by the neural network of the individual brain.

Which is:

$$dU = dU_k + dU_n \tag{5-43}$$

Among them, U_k is the kinetic energy of the social individual, and U_n is the energy consumed by the neural network of the social individual's brain.

For human society, especially in aging society, dU_k can be ignored. In animal societies, dU_n can usually be ignored.

If we change formula (5-41) a bit, we get

$$dQ = dU - dW = dU + pdV \tag{5-44}$$

That is to say, by adding the internal energy of the social thermodynamic system and the external work, the heat change of the entire social thermodynamic system can be obtained.

Since there is an equation of state as a constraint, only two of the three parameters of p, V, and T need to be considered as independent variables in the specific calculation. Then the change in system heat can be specifically calculated.

Math further

If T and V is selected as the independent variable, then

$$dQ = \left(\frac{\partial U}{\partial T}\right)_V dT + \left(\frac{\partial U}{\partial V}\right)_T dV + pdV \quad (5\text{-}45)$$

or

$$dQ = \left(\frac{\partial U}{\partial T}\right)_V dT + \left(\frac{\partial U}{\partial V} + p\right)_T dV \quad (5\text{-}46)$$

If you choose p, T as the independent variable, you can use the following formula:

$$dQ = \left(\frac{\partial U}{\partial T} + p\frac{\partial V}{\partial T}\right)_p dT + \left(\frac{\partial U}{\partial p} + p\frac{\partial V}{\partial p}\right)_T dp \quad (5\text{-}47)$$

If you use p, V as the independent variable, you can use the formula:

$$dQ = \left(\frac{\partial U}{\partial p}\right)_V dp + \left(\frac{\partial U}{\partial V} + p\right)_p dV \quad (5\text{-}48)$$

We can illustrate the characteristics of energy and heat in social thermodynamics with several computational examples.

5.3.2 Changes to internal energy due to work

The work done by external forces on the social system is mainly to cause changes in the urban social space. If the space decreases, the work done will increase the internal energy, otherwise it will decrease the internal energy.

Here

$$dU = dQ + dW \quad (5\text{-}49)$$

If the social space changes, there will be external or internal pressure to work. This work contributes to changes in internal energy.

Which is

$$dU = dQ - pdV \tag{5-50}$$

That is, if social space decreases, the effect of work is to increase internal energy. While social space increases, the effect of work is to reduce internal energy.

Figure 5-1 can be used to represent the amount of work done.

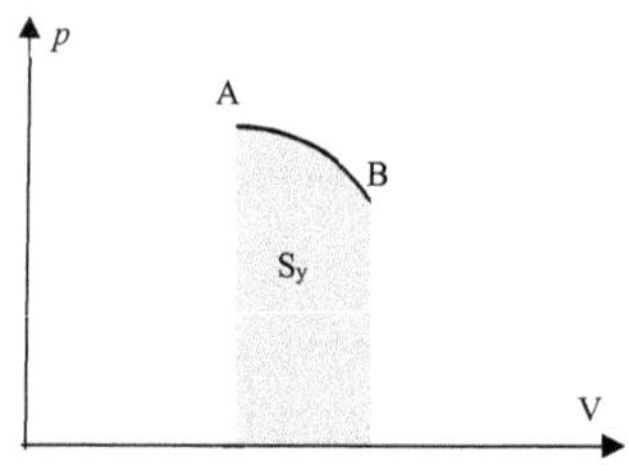

Figure 5.1. The work done by *pV* curve

In the figure, from state A to state B, the AB curve reflects the changes in pressure and volume in the process of doing work. Since the shaded part under the AB curve represents the product of pressure and volume, the shaded part also represents the amount of work done.

Which is

$$dW = -pdV = -dS_y \tag{5-51}$$

Then

$$W = -S_y \tag{5-52}$$

Example 5-6:

For a city with a population of 1 million, it covers an area of 1,000 square kilometers. If the city gradually concentrates its population within a range of 990 square kilometers due to some factors.

Assuming the external pressure is 10,000*Pa*, what is the work done by the calculated pressure? If the original temperature of the city is 272*K*, calculate the temperature after population concentration.

Solution:

Social space is reduced by:

$$dV = -3\times 10\times 10^{6} m^{2} \tag{5-53}$$

$$dW = -pdV \tag{5-54}$$

According to the calculation of a city with a population of 1 million and an area of 1,000 square kilometers, $p \approx 10000Pa$, then

$$dW = 10000\times 3\times 10\times 10^{6} = 3\times 10^{11}\,(J) \tag{5-55}$$

Because this is an open system, we can use Charlie's Law to solve for temperature changes due to volume changes.

$$\frac{V_1}{T_1} = \frac{V_2}{T_2}$$

Considering T_1 =272*K*, so

$$T_2 = \frac{T_1}{V_1}V_2 = \frac{272}{3\times 1000}\times 3\times 990 = 269(K)$$

That is to say, under the condition of equal pressure, the decrease of social space will decrease the temperature.

5.3.3 Thermal capacity and enthalpy

Heat capacity refers to the amount of heat that needs to be absorbed per degree increase in temperature. The formula for calculating heat capacity:

$$C = \frac{dQ}{dT} \tag{5-56}$$

heat Capacity at constant volume

$$C_V = \left(\frac{\partial Q}{\partial T}\right)_V \tag{5-57}$$

Because the volume is unchanged, so $dW = pdV = 0$, so $dU = dQ$.

Considering the ideal situation, internal energy is only a function of temperature, so

$$C_V = \left(\frac{\partial U}{\partial T}\right)_V = \frac{dU}{dT} \tag{5-58}$$

The partial derivative is not used in the formula, mainly considering that U is only a function of T and has nothing to do with p and V.

Then we can calculate the heat Capacity at constant Pressure

$$C_p = \left(\frac{\partial Q}{\partial T}\right)_p \tag{5-59}$$

Then consider: $dQ = dU + pdV$

So the heat Capacity at constant Pressure will formally be:

$$C_p = \left[\frac{\partial(U + pV)}{\partial T}\right]_p \tag{5-60}$$

If we define another thermodynamic parameter

$$H = U + pV \tag{5-61}$$

So the heat Capacity at constant Pressure will be

$$C_p = \left(\frac{\partial H}{\partial T}\right)_p \tag{5-62}$$

where H is called "enthalpy"

Math further

For the ideal gas case, H is only a function of temperature and only varies with temperature, then:

$$C_p(T)=\frac{dH}{dT}=\frac{dU}{dT}+p\frac{dV}{dT} \tag{5-63}$$

Consider ideally:

$$pdV = N\beta dT \tag{5-64}$$

so

$$C_p(T)=C_v(T)+N\beta \tag{5-65}$$

In thermodynamics, heat capacity can only be measured experimentally. However, by establishing a statistical model for the kinetic energy of the microscopic molecular and atom, the principle of the appearance of the heat capacity can be explained from the microscopic level. For substances without quantum effects, if the average kinetic energy of a single molecular is considered, the formula for calculating the heat capacity can be obtained from Boltzmann statistics:

$$C_v = \frac{3}{2}N_n k_B \tag{5-66}$$

where N_n is the number of gas molecules, and k_B is the Boltzmann constant. In applying this formula to social thermodynamic, we can consider several characteristics:

1. The number of macro-social individuals is very small

2. The physical kinetic energy of macro-social individuals is mainly three-dimensional.

3. The energy consumption of social individual brain neural network is two-dimensional.

Therefore, if only one-dimensional is considered, the calculation formula of heat capacity can be expressed as:

$$C_v = \frac{1}{2}N_n k_B \tag{5-67}$$

This is believed to have applications in social systems composed of animals.

If only the energy consumed by the neural network of the brain is considered, the calculation formula of heat capacity can also be expressed as:

$$C_v = N_n k_B \tag{5-68}$$

Since the total number of social individuals is small relative to the microscopic molecular number, we also change the Boltzmann constant to a macroscopic form. Which is:

$$C_v = \frac{1}{2} N\beta$$

Or for three-dimensional

$$C_v = \frac{3}{2} N\beta \tag{5-69}$$

Of course, for two-dimensional case or ease of handling, we can also use directly

$$C_v = N\beta \tag{5-70}$$

For human society

$$\beta \approx 2.76 \times 10^4 \left(J \cdot K^{-1}\right)$$

Example 5-7:

Calculate the heat Capacity at constant volume for a city with a population of 1 million and 1,000 square kilometers under the condition that the urban area remains unchanged. The external pressure is 10,000*Pa*, and the original temperature of the city is 272*K* If the temperature rises by 1*K*, how much does the calculation of heat and pressure need to increase?

Solution:

This involves human physiology and the information processing ability of the neural network of the brain. If we consider that the main heat consumption of human society is in the neural network, according to the statistical mechanics calculation formula of heat capacity (5-70):

$$C_v = N\beta = 10^6 \times 2.76 \times 10^4 = 2.76 \times 10^{10} \left(J \cdot K^{-1}\right) \qquad (5\text{-}71)$$

Then from the definition of heat Capacity at constant volume, we can get:

$$dQ = C_V dT = 2.76 \times 10^{10} dT \qquad (5\text{-}72)$$

If the temperature increases by 1*K*, the heat will increase:

$$dQ = 2.76 \times 10^{10} J \qquad (5\text{-}73)$$

In the case of isovolumics, an increase in temperature causes a change in pressure. This change can be solved according to Guy-Lussac's law:

$$\frac{P_1}{T_1} = \frac{P_2}{T_2}$$

if

$$P_1 = 10000 Pa$$

$$T_1 = 272(K)$$

Then the temperature will increase by 1*K* and the pressure will increase as:

$$P_2 = \frac{P_1}{T_1} T_2 = \frac{10000}{272} \times 273 = 10036(Pa) \qquad (5\text{-}74)$$

Example 5-8:

For a city with a population of 1 million, it covers an area of 1,000 square kilometers. If the average physical temperature of the human body is 310*K*. Find the total internal energy of the thermodynamic system composed of all the population of the city. If the city's annual power generation is 5 billion kilowatt-hours, the average person needs to consume 1,500 kilocalories of heat per day to maintain metabolism. The energy consumed by the neural network of each person's brain and the energy consumed by physiological functions is 50% of the energy consumed by the metabolism. Find the daily social total internal energy change in the city, and calculate the social temperature

Solution:

First, we consider the human body as a physical thermodynamic system, all composed of water molecules. Then the internal energy of the thermodynamic system composed of all N individuals can be calculated by the formula of internal energy:

$$dU_p = C_v' dT \tag{5-75}$$

If we think that a human thermodynamic system is a process from gestation to growth, the system is a process from nothing to something, then its social temperature can be calculated from the beginning of 0*K*. In this way, by integrating the above formula, the internal energy obtained by a human physiological growth can be obtained as

$$U_p = C_v' T = \frac{3}{2} N\beta T = 1.5 \times 10^6 \times 2.76 \times 10^4 \times 310 = 1.3 \times 10^{13} (J) \tag{5-76}$$

Here we use the heat capacity calculation formula of formula (5-69), mainly considering that human physiological activities are carried out in a three-dimensional space.

Then we look at the input energy required to keep the social thermodynamic system running.

The total energy consumed by the city can be calculated by the amount of electricity generated in the city and the average daily calorie consumption per person. The power generation mainly meets the consumption of urban commuting and office logistics. But it is worth noting that what we calculate here is social internal energy, which is the energy consumed by social individuals. In addition to satisfying social individuals to obtain food energy more efficiently, the energy delivered by power plants will be consumed in the physical environment. This physical environment includes thermodynamic systems such as air.

Therefore, only a part of the energy delivered by the power plant is consumed by social individuals, which causes changes in social thermodynamic parameters. The energy consumed by social individuals is limited. It is the energy required for the metabolism of the social individual plus the energy required by the neural network of the social individual and to meet various physiological functions. In this way we can divide the total energy, or total heat, of a society into these two parts:

$$dQ = dE_{nutrition} + \eta \times dE_{nutrition} \quad (5\text{-}77)$$

The η reflects the proportion of energy required to process food, etc. If the degree of industrialization is low, the proportion will be low. If the degree of industrialization is relatively high, the proportion will be high. In this example, $\eta = 50\%$.

Since there are 1 million people in this city, according to the average daily calorie consumption of each person is 1,500 *kcal*, the energy consumed by the entire city to maintain the body's nutritional needs every day is:

$$dE_{nutrition} = 1500 \times 4184 \times 10^6 \approx 6.3 \times 10^{12} (J) \quad (5\text{-}78)$$

Unit conversion above

$$1kcal = 4184J \quad (5\text{-}79)$$

$$1kWh = 3.6 \times 10^6 J \quad (5\text{-}80)$$

In this way, the internal energy change of the city is

$$dU = dQ = 9.45 \times 10^{12} (J) \tag{5-81}$$

It can be seen from formula (5-43) that this internal energy is mainly composed of two parts, namely the physiological kinetic energy dU_k of the social individual and the kinetic energy dU_n of the neural network impulse.

We can also compare the energy delivered by a power plant. Convert annual power generation to daily power generation:

$$dE_{electic} = \frac{5 \times 10^9 \times 3.6 \times 10^6}{365} \approx 5 \times 10^{13} (J) \tag{5-82}$$

It can be seen that the total energy delivered by the power plant is greater than the change in the social internal energy. Most of the excess energy is actually consumed in the atmosphere and causes the physical temperature to rise. This is a direct cause of rising global temperatures.

Finally, we can calculate the total internal energy of the city's thermodynamic system as:

$$U = U_p + dU \approx 2.25 \times 10^{13} (J)$$

For the convenience of calculation, the calculation of social temperature can be based on the formula (5-75), and the heat capacity has been calculated in the formula (5-71). Therefore, the change in social temperature is

$$dT = \frac{dU}{C_v} = \frac{dU}{N\beta} = \frac{9.45 \times 10^{12}}{2.76 \times 10^{10}} = 342(K) \tag{5-83}$$

The use of this form of heat capacity here is mainly due to the fact that most of the nutrients a person absorbs every day are consumed in a two-dimensional brain neural network.

In this calculation, it can be found that $dU_p = 0$. In other words, the basic internal energy to maintain the body's body temperature will not change. So the results calculated by (5-83) reflect changes in social temperature. This is a process of social temperature change in the social system from scratch. because if there is no daily nutrient intake (that is, the continuous input of energy), the entire society does not exist (the social temperature is 0). So, it can be considered that $T = dT$.

In this way, we can get the social temperature of the city as $T = 342K$.

Example 5-9:

Calculate the city's heat Capacity at constant Pressure according to the conditions and calculation results of Example 5-7. Then calculate the change of internal energy.

Solution:

According to formula (5-65), we can get

$$C_p(T) = C_v(T) + N\beta = 2.76\times10^{10} + 10^6\times2.76\times10^4 \qquad (5\text{-}84)$$

So

$$C_p(T) = 5.52\times10^{10}\left(J\cdot K^{-1}\right) \qquad (5\text{-}85)$$

But

$$dQ = C_p dT = 5.52\times10^{10}\times(269-272) = -1.66\times10^{11}(J) \qquad (5\text{-}86)$$

The negative sign reflects that the system will release heat to the outside after the temperature is lowered.

In this way, the internal energy changes to:

$$dU = dQ + dW = -1.66\times10^{11} + 3\times10^{11} = 1.34\times10^{11}(J) \qquad (5\text{-}87)$$

It can be seen that when the social space of the city shrinks, the temperature will drop, and excess heat will naturally be released. However, due to the work done by the pressure, the internal energy of the city still increases.

This occurs in the ideal case of an open system, as opposed to an adiabatic process.

5.3.4 Case study: The impact of Ukrainian wheat on Egypt

In order to make the calculation more intuitive, use integer data as much as possible here. First of all, we can confirm that there are two key data, namely, around 2020, Egypt imports 7 million tons of Ukrainian grain every year, which mainly are wheats. Egypt has a population of about 100 million. In addition, we know that heats in wheat have about 300 *kcal* per 100 grams. Secondly, we convert some of the data into the International System of Units, namely, Egypt imports $7 \times 10^9\, kg$ of Ukrainian grain, and the energy provided by each $1kg$ of wheat is about $1.3 \times 10^7\, J$.

So the energy that these imported wheats can supply to Egypt each year is:

$$dE_{wheat} = 7 \times 10^9 \times 1.3 \times 10^7 \approx 9 \times 10^{16}\,(J)$$

According to the average daily calorie consumption of each person is 1500 kcal, for a population of 100 million in Egypt, the nutrients required to maintain daily neogenesis are:

$$dE_{nutrition} = 1500 \times 4184 \times 10^8 \approx 6.3 \times 10^{14}\,(J) \tag{5-88}$$

From here it can be seen

$$dE_{wheat} > dE_{nutrition}$$

Even taking into account that humans are not able to fully absorb all the energy of food, and that imported food can be further processed, the food imported from Ukraine is sufficient to meet the nutritional needs of the entire population of Egypt. This also shows that the entire country of Egypt is basically completely dependent on the food imported from Ukraine. If the import of Ukrainian wheats is suddenly interrupted, and alternative food cannot be found immediately, it will have disastrous consequences for the entire country.

We can also analyze the impact of these imported wheats on social stability by calculating the social temperature in Egypt. Of course, in addition to absorbing energy from food, society as a whole also needs to obtain energy from other sources. This energy is mainly provided by power plants. These energies include heating food, stabilizing ambient temperature, and more. These energies will also be converted into the internal energy of the whole society. However, these energies cannot be completely absorbed by the human body. Depending on the level of industrialization, the ratio of these external energies converted into the internal energy of society will also vary. It is estimated here that each person can increase calorie consumption by about 50% due to the input of these external energy. So the internal energy of the whole society will be:

$$dU = dE_{nutrition} + 0.5dE_{nutrition} \approx 1 \times 10^{15}(J) \tag{5-89}$$

In this way, the temperature of the whole society can be calculated

$$dT = \frac{dU}{C_v} = \frac{dU}{N\beta} = \frac{1 \times 10^{15}}{10^8 \times 2.76 \times 10^4} \approx 360(K) \tag{5-90}$$

As Examples 5-8 illustrate, this is the social temperature of the society. This temperature is of course higher than the ambient temperature, which is also the result of industrialization. After all, the result of industrialization is that it can provide society with more sufficient energy input, and then obtain additional energy from the environment to meet the needs of society. Of course, if there is no industrialization, on the one hand, nutrition will be reduced, and other energy used to meet the needs of the human body will also be reduced. This can effectively reduce the social temperature.

Of course if Egypt's food imports from Ukraine are interrupted. It means that the energy obtained by each social individual will decrease. Of course, if it is lower than the normal human body needs, such as the energy intake of each person is less than 1,000 kcal per day, the human body will be in a state of starvation, which will bring about a series of more serious social problems. Of course, due to globalization, it is still possible to find alternative foods. It's just that these alternative foods may not have the same energy as Ukrainian food. This results in a reduction in the amount of energy obtained by each social individual to 1,200 kcal, and other energies remain unchanged, then

$$dE'_{nutrition} = 1200 \times 4184 \times 10^8 \approx 5 \times 10^{14}\,(J)$$

The internal energy of the whole society will become

$$dU = dE'_{nutrition} + 0.5dE_{nutrition} \approx 8 \times 10^{14}\,(J) \tag{5-91}$$

The social temperature will become

$$dT = \frac{dU}{N\beta} = \frac{8 \times 10^{14}}{10^8 \times 2.76 \times 10^4} \approx 290(K)$$

This temperature appears to be close to the ambient temperature. From the perspective of the 0th law of thermodynamics, it is more helpful for society to absorb energy from the environment. But there is a precondition here, which is that Egypt must have access to enough alternative food. At the same time, some mandatory food distribution systems are required, such as food ration cards. Only in this way can all the nationals be equally distributed to all the food and ensure that there is no local or regional famine. Of course, this mandatory distribution system also goes against the current trend of democratization and globalization, so it is not a good solution to the problem. From the perspective of the development history of human society, the more serious problem is that if a society changes from one system to another suddenly, it is likely to bring about violent social unrest.

5.3.5 Reversible and adiabatic processes

The process would be reversible if there was no loss of friction and inter-molecular forces in the thermodynamic process. The actual social thermodynamic systems are systems far from the equilibrium state, and energy loss is unavoidable in the process of system operation.

However, when calculating, for a process that occurs in a relatively short time, if the loss of energy is small, it can be approximately regarded as reversible. For example, when working in a city, a large number of social individuals commute from the suburbs to work in the urban area. At this time, people will be concentrated in the urban area. When it is time to get off work, a large number of social individuals will return to the suburbs from the urban area. In this case, there will be short-term changes in social space. Because the same process is repeated every day. The energy loss caused by social pressure, temperature and space changes in this process is relatively small, so it can be approximately regarded as a reversible process.

It can be seen from Figure 5-1 that a system changes from the state of point A to the state of point B through the thermodynamic process, due to the increase in volume, $dV > 0$ and pressure $p > 0$, the system will do negative work. But if the other way around, along the same curve, the system changes from point B to point A, the system will do positive work due to the volume reduction, $dV < 0$. In the figure, the area under the curve AB is the same no matter which direction it goes along, and the negative and positive work can cancel each other out. This means that from state A to state B, and then from state B back to state A along the same curve, this is a reversible thermodynamic process.

But Figure 5-2 shows a different process, where the system does negative work from point A, along point B, and finally to point C. But to get back from point C to point A, you have to go through point D. At this time, the system is doing positive work. However, the area under the ABC curve is significantly larger than the area under the CDA curve. The difference between the two is the area S_y of the shaded part in the figure. This shows that the system state does external work when it changes. And this kind of work will cause the change of system energy. Changes in energy will cause changes in the system's internal energy and heat. That is to say, after the cycle of ABCD, although the parameters such as pressure and volume of the system have returned to the starting point, the internal energy, heat and temperature of the system have undergone irreversible changes. This is an irreversible process.

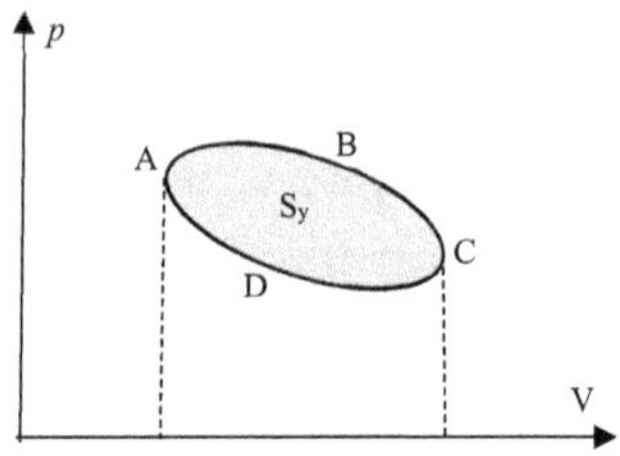

Figure 5.2. Thermodynamic cycle

In addition to the reversibility of the process, many other thermodynamic systems will also have the process of exchanging heat with the outside world. However, If the system cannot exchange heat with the outside world, the process is an adiabatic process. For an adiabatic process, it means that there is no heat exchange between the system and the outside world. At this moment

$$dQ = 0$$

This is the basic condition of the adiabatic state.

Math further

Atthis time, the formula can be obtained:

$$dU = dQ + dW = dW = -pdV \tag{5-92}$$

At this time, because there is no heat transfer, it also means that the calculation can only be performed according to formula (5-58), namely:

$$dU = C_v dT = -pdV \tag{5-93}$$

Then for the ideal gas equation of state (5-7), differentiating both sides can get:

$$pdV + Vdp = N\beta dT \tag{5-94}$$

Dividing both sides by pV, we get:

$$\frac{dp}{p} + \frac{dV}{V} = \frac{dT}{T} \tag{5-95}$$

Due to the constraints of formula (5-93), it can be found that dT is not an independent variable, so by substituting dT of formula (5-93) into formula (5-95), we can get:

$$\frac{dp}{p}+\gamma\frac{dV}{V}=0 \tag{5-96}$$

Among them

$$\gamma=\frac{C_V+N\beta}{C_V}=\frac{C_p}{C_V}=Const. \tag{5-97}$$

The γ is called the adiabatic coefficient.

Solving equation (5-96), the equation of state for the adiabatic process can be obtained as:

$$pV^{\gamma}=Const. \tag{5-98}$$

Combining the equations of state for ideal gases (5-7), we can also eliminate p and V to obtain two other adiabatic equations:

$$TV^{\gamma-1}=Const. \tag{5-99}$$

$$p^{\gamma-1}T^{-\gamma}=Const. \tag{5-100}$$

Equation (5-98) is actually a hyperbolic equation. That is to say, the adiabatic line is a hyperbola on the *pV* coordinate. So we can see that the Figure 5-1 is not a adiabatic line.

Example 5-10.

For a city with a population of 1 million and an area of 1000 square kilometers, the external pressure is 10,000*Pa*. The social temperature is 272*K*. If the information pressure caused by the need to work is considered, which will cause social space of the city reduced to 900 square kilometers in normal working hours during the day, and the social space of the city will be restored to 1,000 square kilometers after get off work. The height of the social space of the whole city is 3 meters, then try to calculate the changes in social pressure and social temperature in the process of commuting to work throughout the city.

Solution:

Since the commuting process of a day is very short relative to the change process of the whole society, this process can be regarded as an adiabatic process.

First, from the heat capacity calculated in Example 5-7 and Example 5-9, we can calculate the adiabatic coefficient

$$\gamma = \frac{C_p}{C_V} = \frac{5.52 \times 10^{10}}{2.76 \times 10^{10}} = 2 \quad (5\text{-}101)$$

So the adiabatic equation of state is:

$$pV^2 = A \quad (5\text{-}102)$$

Here we first calculate the constant C. Calculated according to the initial state $V = 3 \times 10^9 m^3$, according to:

$$p_1 = 10000Pa$$

but

$$A = p_1 V_1^2 = 10000 \times \left(3 \times 10^9\right)^2 = 9 \times 10^{22} \left(Pa \cdot m^3\right) \quad (5\text{-}103)$$

Therefore, if the social space of the whole city is reduced to 900 square kilometers due to the additional information pressure after going to work, then

$$p_2 \left(3 \times 9 \times 10^8\right)^2 = 9 \times 10^{22} \quad (5\text{-}104)$$

We can get

$$p_2 \approx 1.24 \times 10^4 \left(Pa\right) \quad (5\text{-}105)$$

Of course, we can also directly use

$$p_1V_1^{\ 2} = p_2V_2^{\ 2} \tag{5-106}$$

get

$$p_2 = \frac{p_1V_1^{\ 2}}{V_2^{\ 2}} \approx 1.24 \times 10^4\,(Pa) \tag{5-107}$$

The two results are consistent.
Then by formula (5-99), we can also get:

$$T_1V_1^{\ 2-1} = T_2V_2^{\ 2-1} \tag{5-108}$$

$$T_2 = \frac{T_1V_1}{V_2} = \frac{272 \times 3 \times 10^9}{3 \times 9 \times 10^8} \approx 302K \tag{5-109}$$

It can be seen from the calculation results that the social temperature has risen during work hours. However, this temperature may have exceeded the ambient temperature. So it becomes a bit difficult to get the energy in the environment. However, in the era of industrialization, humans can obtain energy from the environment by consuming additional energy to meet their own nutritional needs. This extra energy usually means that the energy consumed by the city during the day at work exceeds the energy consumed by the entire city at night.

In addition, for an adiabatic process, we can also see that when social space decreases, social pressure and temperature both increase to a certain extent. This is the exact opposite of what the open system shown in Example 5-6 would have calculated. This is due to the fact that an open system can release heat or absorb heat out of the system, resulting in a drop in temperature as the volume is compressed.

We can also compare this result with the isothermal process of Example 5-3. The result of both examples is that the social space decreases, resulting in an increase in pressure, but for the compression ratio of social space in Example 5-10, if it is an isothermal process, the pressure is proportional to the change in social space. The increase in social pressure is proportional to the cubic multiple of the increase in social space in the adiabatic process. In addition, the adiabatic process has the same change trend compared with the isobaric process.

5.3.6 Carnot Cycle

If a thermodynamic state starts from the starting point, the pressure and volume are constantly changing, and finally return to the starting position. This process is a thermodynamic loop.

As can be seen from Figure 5-2, the starting point is point A. When the thermodynamic process starts from point A and runs along the BCD direction, and finally returns to point A, this is a clockwise process. Since the volume increases continuously during the ABC process, the pressure work is negative work and needs to absorb heat. The thermodynamic process goes from point C through point D to point A, and the volume is reduced at this time. So pressure does positive work, and the process releases heat. But the area corresponding to this positive work is obviously smaller than the area under the ABC curve. Therefore, the cycle running clockwise along the ABCD direction does negative work, and the entire cycle process will continue to absorb heat from the environment. Of course the thermodynamic system absorbs energy from the environment, which will cause the temperature of the system to rise.

A counterclockwise cycle, in turn, it will release energy into the environment. The process by which the system releases energy to the environment means that the temperature of the system will drop. This is the principle of refrigeration and air conditioning.

However, in order to efficiently output heat or perform refrigeration, a special Carnot cycle is required. A Carnot cycle consists of two quasi-static isothermal processes and two quasi-static adiabatic processes. As shown in Figure 5-3.

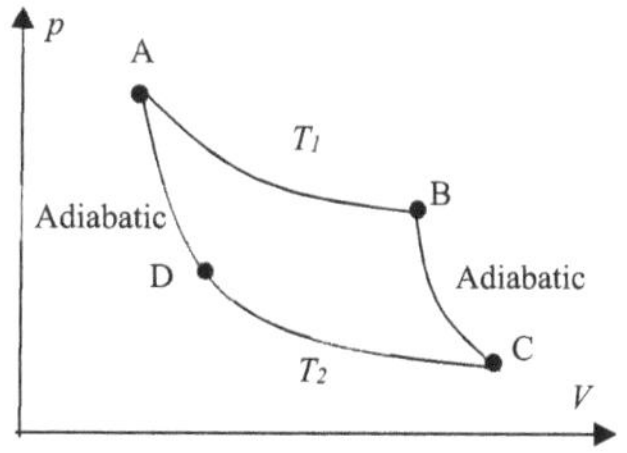

Figure 5.3. Carnot cycle

In Figure 5-3, going from A to B is an isothermal process, and this process results in a decrease in pressure as the volume increases due to the increase in volume. Since it is doing negative work, the cycle absorbs heat from the outside. B to C is an adiabatic process. During this process, due to the increase in volume, the temperature rapidly drops to T_2. Then it is an isothermal process from C to D. At this point the temperature remains the same at T_2, but the volume begins to decrease. The process of volume reduction means that the pressure starts to do positive work, which will release heat to the outside. From D to A is another adiabatic process. The volume of the process is reduced, so the pressure is also doing positive work. But since it is an adiabatic process, all positive work is converted into the internal energy of the system in order to raise the temperature to T_1. Since the system does more negative work than positive work, it means that the system absorbs more heat from the outside than it releases. This also means that the Carnot cycle is able to continuously generate heat. This is called a heater. And if the reverse is a counterclockwise cycle, the system will be able to continuously release heat to the outside, resulting in a cooling effect. This is the cooler.

It can be seen from the Carnot cycle that although the internal temperature of the cycle will continue to change, this constantly changing process can always oscillate within a certain range, which is why the Carnot cycle can continue. Of course, to achieve such a cycle, whether it is a heater or a cooler, additional energy is required to perform work to maintain it.

5.3.7 Case study: Holiday effect

Generally speaking, every country in the world has a relatively long public holiday every year. During public holidays, almost everyone can be on vacation. At this time, the urban population was quickly evacuated to the surrounding areas. When the holiday is over, all the people on holiday will quickly converge on the city to work.

This holiday time can be divided into four stages: the first stage is the start of the holiday, the rapid evacuation of personnel, and the rapid increase of urban social space. The second stage is when the evacuation is completed and the vacation begins. At this time, the social space remains unchanged and is in an isovolume process. The third stage is when the holidays are coming to an end and people start returning to the city quickly. The social space is shrinking rapidly. In the fourth stage, once people return to work in cities, the social space will not change at this time. This is again an isocratic process.

The whole process can be represented by Figure 5-4.

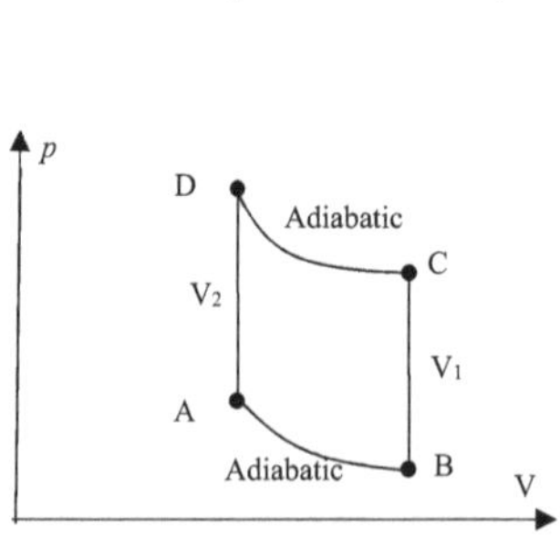

Figure 5.4. Holiday cycle

As can be seen from Figure 5-4, A is the starting point and the time point when the holiday starts. The evacuation of people during the holiday led to the rapid expansion of the social space. Since this expansion time was very short, it was an adiabatic process. The graph corresponds to the curve from point A to point B, there is no heat input and output on the adiabatic curve AB. We can notice that as the volume expands, the pressure of the adiabatic process also decreases rapidly, and a reduction in pressure corresponds to a sharp drop in temperature. So what pressure does in this process is negative work. Considering that there is no heat output during the adiabatic process, this work actually leads to the reduction of the internal energy of the entire social system. Therefore, when it comes to point B, the temperature of the entire social system has dropped to a relatively low level. And once enough people are evacuated, all the spaces of everyone will begin to stabilize. Going from point B to point C is an isovolumic process. This is the time when everyone starts to go on vacation. At this stage, people start to consume a lot. As the consumption process is affected by various information interactions, internal pressure begins to rise. For an isovolumic process, an increase in pressure causes an increase in social temperature. The rise of social temperature leads to a decrease in the difference between social temperature and the physical temperature of the environment, thereby inhibiting the society from acquiring energy from the outside, that is, inhibiting consumption. Therefore, in the BC stage, an important sign is that the consumption amount is a decline process from high to low.

Curve CD is also an adiabatic curve. This is after the holidays are over and the holiday crowds are starting to return to the city. Since this time is also relatively short, it can also be regarded as an adiabatic process. Although the pressure does work, the CD stage also has no heat input and output. After the CD stage is over, the social temperature will rise further, and of course, consumption will be further suppressed. After reaching point D, it means that all the people have returned to the city and returned to their jobs. At this time, the social space does not change any more, so DA is also an isovolumic process. Since point A is less pressure than point D, it is a process of constant release of social pressure. No change in the social space also means that the pressure is not doing the work. The drop in temperature is all caused by the drop in internal energy. The drop in temperature also means that consumption is starting to gradually return to normal.

We can verify this by comparing some important holiday spending changes. One of the most important long holidays in China is the Spring Festival and the National Day holiday. The Chinese long holiday is more typical. Because there are no other holidays that conflict with each other during this period. This can better reflect the changes in social thermodynamic parameters of the 7-day long vacation after the intense work in the urban society.

The consumption patterns of long holiday are mainly tourism and cultural entertainment. Among all consumption, the most sensitive to change is the change of the theater box office. Considering that the viewing time is only one or two hours, the box office consumption of the theater can more quickly reflect the change of consumption willingness. Consumption such as tourism requires a long time and a more cumbersome preparation process, so the reflection of consumption willingness will be delayed.

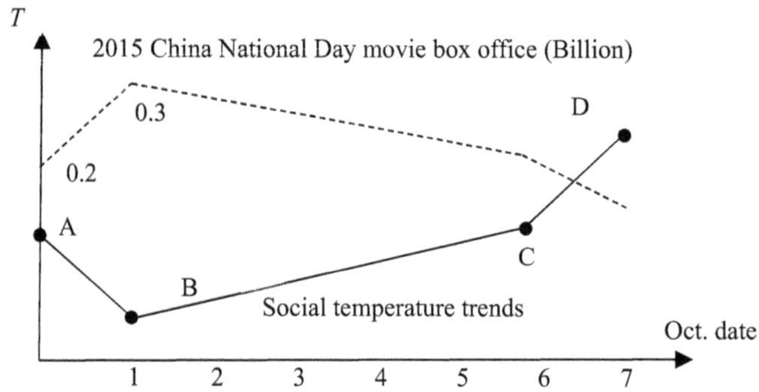

Figure 5.5. Comparison of social temperature change and consumption

It can be seen from Figure 5-5 that the trend of social temperature changes is reflected in the discounted ABCD. The social temperature decreases from point A to point B through adiabatic expansion. Due to the decrease of social temperature, it is easier for external energy to be input into the system, so it is theoretically predicted that there will be a rapid increase in consumption during this period. Then from October 1st, through the isovolume process from point B, the social temperature starts to rise steadily to point C. The rise of social temperature means that the ability of the system to obtain energy from the environment begins to decline, indicating that there will be a process of continuous decline in consumption during this period. On October 6th, the long holiday was coming to an end, and the crowd began to return to the city, which was an adiabatic compression process. This causes social temperature to rise rapidly from point C to point D. The rapid rise in temperature also means that there will be a process of rapid decline in consumption.

Above the social temperature change line is the dotted line of the change trend of the Chinese film box office during the National Day in 2015. From this straight line, we can see that the box office of the movie has risen from about 200 million yuan before October 1st to about 300 million yuan on October 1st. Then the box office revenue steadily declined from No. 1 onwards. On the 6th, the box office basically recovered to about 200 million yuan. Then, with the rapid increase of social temperature, the box office revenue has a rapid decline process, and by the 7th, the box office revenue has decreased even before the National Day.

In fact, there is an inverse relationship between changes in consumption during the National Day and changes in social temperature. In a year, we can also find that the trend of Chinese movie box office is basically the same. That is, when the Spring Festival holiday begins, the movie box office consumption will rise rapidly, and then it will decline steadily. After the long holiday, the box office revenue of the movie began to rise steadily until July, the arrival of the summer vacation. At the beginning of the summer vacation, the students in the school quickly returned home from all levels of schools, which meant that the students' social space expanded adiabatically. Considering that students are the mainstay of the movie box office, the movie box office revenue will rise rapidly at the beginning of the summer vacation. After that, it continued to decline until the beginning of the National Day.

From the conversion relationship between heat energy and work, the adiabatic process of process AB in Figure 5-4 is due to the expansion of social space. According to formula (5-92), it can be seen that this process is doing negative work, that is, $W_{AB} < 0$.

Because the social space did not change, the BC process did not work regardless of the change in pressure. However, due to the increased pressure of the BC process, it means that heat is absorbed from the outside. This is an endothermic process, that is, $Q_{BC} > 0$. Then $W_{CD} > 0$. $Q_{DA} < 0$.

Considering that the area under the CD adiabatic line is larger, $W_{CD} + W_{AB} > 0$. That is to say, this process is the work done by the external environment on the social system. According to the conservation of energy, it means that the social system will convert this work into heat and release it. That is, $Q_{BC} + Q_{DA} < 0$. That is, the heat absorbed by the system is greater than the heat released by the system.

If the above cycle process is rotated in a counterclockwise direction, it can be used to correspond to the daily commuting process of the urban population. First start from point B and go through the adiabatic process to point A. This process corresponds to the process of going to work. The urban social space is rapidly reduced, while the work pressure is rapidly increasing, and the social temperature is rapidly rising. Then, the isovolume process from point A to point D corresponds to the work period. In this process, work pressure continues to increase, resulting in a continuous rise in social temperature. The adiabatic process from point D to point C corresponds to the off-duty time. During this period, the social space expanded rapidly and the social temperature decreased rapidly. From point C to point B, most of the population is resting at home, which is also an isovolume process, and work pressure has continued to drop, and social temperature has also dropped proportionally.

Of course, if the cycle process in Figure 5-4 is in a counterclockwise direction, it is the "Otto cycle" commonly used in engines. Using this cycle, the heat of energy such as gasoline can be converted into the power output of the engine.

5.4 The second law of thermodynamics

5.4.1 The formulation of the second law of thermodynamics

The second law of thermodynamics reflects how heat is converted into "work". Because in human society, although heat is directly required in many fields, work is also required in order to obtain kinetic energy similar to that produced by a heat engine.

Of course, the ideal situation is that when heat is needed, we can completely convert work into heat at any time. When work is required, the heat can be completely converted into "work" at any time. However, the second law of thermodynamic tells us that although to successfully convert heat or convert work into heat, this process can be completed by different thermodynamic cycles, such as Carnot cycle, etc., but the conversion efficiency between heat and work is impossible to reach 100%. Therefore each conversion process consumes additional energy. Just like a battery, the amount of electricity charged must be greater than the amount of electricity released. One of the losses is that the battery heats up while charging.

There are two expressions of the second law of thermodynamics, among which the Kelvin Statement reflects that it is impossible to obtain heat from a single heat source and convert it successfully without consuming external energy. The Clausius Statement means that it is impossible to transfer heat from a low temperature system to a high temperature system without consuming external energy. The two representations are essentially the same, that is, they both reflect the irreversibility of thermodynamics and social systems.

Although the social system is a system far from equilibrium, if we look at the social thermodynamic system from a macro perspective, we will find that a social thermodynamic system is actually composed of a large number of social individuals. The interaction between these social individuals is very similar to the molecular interaction between gases, liquids and even solids, and the social thermodynamic system also has different states. Therefore, from a more macro perspective, we can also apply some important laws of thermodynamics to the social thermodynamic system.

Of course, applying the laws of thermodynamics to the social thermodynamic system, we should also notice that in the social thermodynamic system, the interaction between social individuals is very strong. Therefore, the condensed state formed in the social thermodynamic system is also very strong. These are also some issues that should be paid attention to when some important laws of thermodynamics are applied to the social thermodynamic system, without take into account the peculiarities of the social system, very absurd conclusions can be drawn.

Therefore, in the following application processes, we all use an approximate method. That is to say, within certain conditions and ranges, these thermodynamic laws can be approximately established in the social thermodynamic system. Once the constraints of these conditions are left, then completely different results may be obtained.

5.4.2 Entropy

To represent this irreversibility of thermodynamic systems, Clausius proposed the concept of entropy. This important function used to represent the state of the system in the thermodynamic system is *S*

$$dS = \frac{dQ}{T} \tag{5-110}$$

Interestingly, the above definition requires that the thermodynamic system must be reversible. Since many systems are all reversible considering the microscopic processes in infinitely small time scales. Therefore, the above conditions can be approximately established in the infinitely small process.

Math further

For macroscopic processes, there are multiple paths due to thermodynamic loops. So integration along different paths leads to different results. After analysis, Clausius found that

$$\Delta S \geq \int_{p_1}^{p_2} \frac{dQ}{T} \tag{5-111}$$

Now we can also use entropy S for heat:

$$dQ = TdS \tag{5-112}$$

This is our common relationship between heat or energy and entropy.

From this we can also express the entropy function with other state parameters of thermodynamics.

According to the formula:

$$dQ = TdS = dU + pdV \tag{5-113}$$

You can get

$$dS = \frac{\partial S}{\partial U}dU + \frac{\partial S}{\partial V}dV = \frac{dU}{T} + p\frac{dV}{T} \tag{5-114}$$

so that you can get

$$\frac{\partial S}{\partial U} = \frac{1}{T} \tag{5-115}$$

$$\frac{\partial S}{\partial V} = \frac{p}{T} \tag{5-116}$$

However, it is difficult to calculate entropy using thermodynamic macro parameters. Therefore, we mainly study the microscopic number of states, and then use Boltzmann's definition of entropy to calculate the entropy of the thermodynamic system, and then obtain other state parameters of the thermodynamic system from the entropy obtained from the microscopic.

Boltzmann entropy is defined as:

$$S = -\sum_i p_i lnp_i \tag{5-117}$$

where p_i is the probability of being in a certain microstate. However, in statistical physics, p_i can already be converted into a microscopic number of states W that calculates a distribution, so that the Boltzmann entropy can be formally expressed as:

$$S = k_B lnW \tag{5-118}$$

Therefore, the method to obtain the entropy with the knowledge of microscopic statistical physics is to obtain the microscopic number of states *W* of the thermodynamic system. To solve this problem, the so-called "ensemble" theory is mainly used in statistical mechanics. The canonical ensemble deals with the situation where a thermodynamic system is in contact with a very large heat source. This situation is basically consistent with the social thermodynamic system. In fact, an isolated social thermodynamic system is almost non-existent. After all, a society can form and maintain stability because of a near-infinite supply of energy. And this near-infinite energy supply relies mainly on solar energy. So solar energy can be seen as this very large heat source.

In addition, quantum effects do not need to be considered when dealing with social thermodynamic system problems. Therefore, the partition function of the classical ideal gas can be approximated to obtain the microscopic number of states of the social thermodynamic system.

However, as we have mentioned many times before, the social thermodynamic system is more macroscopic than the physical thermodynamic system, so some microscopic parameters, such as the Boltzmann constant, need to be replaced by a more macroscopic constant. To this end, let's first look at the entropy formula solved by statistical physics for an ideal gas (Pathria & Beale, 2011):

$$S = \frac{3}{2}N_n k_B lnT + N_n k_B ln\frac{V}{N_n} + \frac{3}{2}N_n k_B\left\{\frac{5}{3} + ln\left(\frac{2\pi m k_B}{h^2}\right)\right\} \qquad (5\text{-}119)$$

In this formula, k_B is the Boltzmann constant, h is Planck's constant, and m is the molecular mass. N_n is the total number of molecules, V is the volume, and T is the temperature.

It can be seen that, according to the previous transformation method of Boltzmann constant, we can use

$$\beta = n_p k_B \qquad (5\text{-}8)$$

instead of the Boltzmann constant.

For another microscopic term containing molecular mass, Boltzmann constant and Planck constant, we can also transform it, so that we can get:

$$\frac{2\pi mk_B}{h^2}=\frac{2\pi(n_pm)(n_pk_B)}{(n_ph)^2}=\frac{2\pi M\beta}{\mathfrak{H}^2} \tag{5-120}$$

In this way, M can represent the mass of the human body, β is the macroscopic Boltzmann constant, and $\mathfrak{H}$ is the macroscopic Planck constant. However, the mass of the water molecular and the Boltzmann constant and Planck constant can be used directly in the calculation.

After the above transformation, the formula of entropy becomes:

$$S=\frac{3}{2}N\beta lnT+N\beta ln\frac{V}{Nn_p}+\frac{3}{2}N\beta\left[\frac{5}{3}+ln\left(\frac{2\pi mk_B}{h^2}\right)\right] \tag{5-121}$$

In the above formula, N is the number of social individuals in the society, $n_p=2\times10^{27}$ is the number of water molecules contained in each social individual. The value of each parameter is

For human society, $\beta=2.76\times10^4\,J\cdot K^{-1}$

Boltzmann constant, $k_B=1.38\times10^{-23}\,J\cdot K^{-1}$

Planck constant, $h=6.63\times10^{-34}\,J\cdot s$

Water's molecular mass： $m=3\times10^{-26}\,kg$

This results in one of the fixed numerical terms:

$$ln\left(\frac{2\pi mk_B}{h^2}\right)=ln\left(\frac{2\pi\times3\times10^{-26}\times1.38\times10^{-23}}{6.63^2\times10^{-68}}\right)\approx43 \tag{5-122}$$

This makes it possible to correlate microstates with macroscopic thermodynamic parameters. The whole solution process is to first obtain the entropy from the canonical ensemble, and then obtain other thermodynamic parameters from the thermodynamic equation.

Example 5-11.

For a city with a population of 1 million and an area of 1,000 square kilometers, the social temperature is 272*K*. Calculate the entropy of the city.

Solution:

Substituting the above parameters into formula (5-121), we can get:

$$S = 10^6 \times 2.76 \times 10^4 \times \left\{ \frac{3}{2} ln272 + ln \frac{3 \times 10^9}{10^6 \times 2 \times 10^{27}} + \frac{3}{2} \times \left[\frac{5}{3} + ln \left(\frac{2\pi m k_B}{h^2} \right) \right] \right\} \tag{5-123}$$

Finally get

$$S \approx 5.8 \times 10^{11} \left(J \cdot K^{-1} \right) \tag{5-124}$$

5.4.3 Entropy increase of balanced system

If there is a dense enough mass of gas in a vacuum, that gas will expand freely to become a very rarefied gas. "expand freely" means that the gas is not under any pressure. The expansion process of the gas is also adiabatic due to the expansion in a vacuum. So:

$$dQ = 0 \tag{5-125}$$

In addition, although the volume increases, because it is an ideal gas, the pressure is 0

which is

$$dW = pdV = 0 \tag{5-126}$$

Since there is no work done and no heat change, there is no change in the temperature of the gas.

Because this expansion time is long enough, we can integrate it according to a certain path, and we can also get

$$\int_{V_1}^{V_2} \frac{dQ}{T} = 0 \tag{5-127}$$

However, according to the entropy calculation formula (5-121) we obtained, the volume of the gas expands, and the change in entropy will not be 0:

which is

$$\Delta S = N\beta\left(ln\frac{V_2}{Nn_p} - ln\frac{V_1}{Nn_p}\right) = N\beta ln\frac{V_2}{V_1} \tag{5-128}$$

because

$$V_2 > V_1 \tag{5-129}$$

therefore

$$\Delta S > \int_{V_1}^{V_2} \frac{dQ}{T} \equiv 0 \tag{5-130}$$

That is to say, the entire system has an increase in entropy without absorbing heat and external work.

Considering that the gas expands freely in a vacuum, there is only one direction, that is, the volume increases. Therefore, the entropy of the system is always developing in a larger and larger direction. This is also an example of entropy increase.

Of course if the gas is compressed adiabatically in a vacuum, the entropy will decrease. This entropy decrease process is due to external force work. That is, by consuming external energy, the entropy of the system is reduced. This does not contradict the second law of thermodynamics.

Entropy increase in social systems is a more complex process because social systems are far from equilibrium. However, if the interaction between social individuals is relatively weak, this may lead to easier evacuation between social members. For example, the free diffusion of a city's population may lead to an increase in entropy. Of course, such an entropy increase process cannot be completely free from changes in energy. For example, since it involves the interaction between social individuals within the system, this means$p_i \neq 0$, which means:

$$dW = -p_i \Delta V < 0 \tag{5-131}$$

That is to say, in the process of free adiabatic expansion, the social thermodynamic system will do external work. During this process, entropy is still increasing. In the end, when the social space no longer increases, the system no longer does external work, and the entropy is also maintained at a maximum level.

5.4.4 Free energy

The second law of thermodynamics can be expressed as:

$$\Delta S \geq \int_1^2 \frac{dQ}{T} \tag{5-132}$$

If the temperature remains constant, then

$$T\Delta S \geq Q \equiv \Delta U - W \tag{5-133}$$

The above formula is transformed after integrating the differential form of the first law of thermodynamics.

So we get

$$\Delta U - T\Delta S \leq W \tag{5-134}$$

or

$$(U_2 - TS_2) - (U_1 - TS_1) \leq W \tag{5-135}$$

If we define a new function:

$$F = U - TS \tag{5-136}$$

but

$$\Delta F \le W \tag{5-137}$$

It can be seen that F is also a form of energy. This is called Helmholtz free energy.

Considering the meaning of entropy S, the equal sign here means that this is a reversible isothermal process. The less than sign means that this is an irreversible isothermal process.

The above is an isothermal process, and we can define a free energy F. But if it is an isobaric process, the expansion work can also be considered, and we can also define a new function:

$$G = F + pV \tag{5-138}$$

Then the function is called the Gibbs function, also known as Gibbs free energy.

Helmholtz free energy and Gibbs free energy, like entropy, are state parameters of a social thermodynamic system. In actual use, flexible selection is mainly made according to the actual thermodynamic conditions.

References

Pathria, R. K., & Beale, P. D. (2011). *Statistical mechanics*. Elsevier Science. The Boulevard, Langford Lane, Kidlington, Oxford, 0X5 1GB, UK: 1–23

Van Ballenberghe, V., Erickson, A. W., & Byman, D. (1975). Ecology of the timber wolf in northeastern Minnesota. *Wildlife Monographs*, (43), 3–43.

6

Social thermodynamic phase transitions

6.1 Changes in the number of social individuals

6.1.1 Social thermodynamic system with variable number of individuals

Like most thermodynamic systems, social thermodynamic systems also have a large number of social individuals that can be changed at will. For example, in a city, there is a large number of individual mobility problems. A large number of people will move from the countryside to the cities. And for cities, there is a large inflow and outflow of people to other cities.

This change in the number of social individuals will directly lead to changes in the interaction between social individuals. Judging from our previous definitions of interaction, with the change of the number of social individuals, the strength of the interaction between the social individuals will also change. Changes in the strength of interactions between individuals will directly change the internal pressure in the social thermodynamic system. This internal pressure is defined from the Van der Waals equation. Changes in internal pressure also led to changes in other thermodynamic parameters and various free energies of the social system.

Changes in thermodynamic parameters lead to changes in the entire thermodynamic system, which in turn lead to social changes.

In thermodynamics, the change of the number of particles usually causes the change of the thermodynamic chemical potential, that is, by multiplying the chemical potential by the change of the number of particles, we get the change of the thermodynamic internal energy. The change in energy caused by this change in the number of particles can be expressed as:

$$\mu dN$$

where μ is the chemical potential, and dN is the change in the number of particles or social individuals. In this way, in addition to using social pressure, social space and social temperature to describe a social thermodynamic system, we can also use chemical potential to describe the overall change of the thermodynamic system caused by the change of particle number.

Of course, in social thermodynamics, we prefer to use macroscopic parameters, that is, N to represent the number of social individuals. In this way, the chemical potential also needs to be multiplied by a coefficient, which is the number of water molecules and so on contained in each social individual n_p.

6.1.2 Variable particle number thermodynamic equations

The aforementioned Helmholtz free energy and Gibbs free energy have important application values in thermodynamic equilibrium and non-equilibrium states.

Math further

Helmholtz free energy and Gibbs free energy can also be expressed in differential terms. When converting these two free energies into differential methods, the main thing is to pay attention to how each parameter change causes the free energy change.

First of all, Helmholtz free energy is defined in the case of constant pressure, so its differential form can be expressed as.

$$F = U - TS \tag{6-1}$$

considering.

$$dQ = TdS = dU + pdV \tag{6-2}$$

If it is an adiabatic process, there is no change in heat, so it can be obtained by differentiating both sides of the Helmholtz free energy.

$$dF = -SdT - pdV \tag{6-3}$$

This way we can also obtain functions of entropy and pressure.

$$S(T,V) = \left(\frac{\partial F}{\partial T}\right)_V \tag{6-4}$$

$$p(T,V) = -\left(\frac{\partial F}{\partial V}\right)_T \tag{6-5}$$

For Gibbs free energy, the same method can be used to obtain its differential form.

$$dQ = TdS = dU + pdV \tag{6-6}$$

$$G = F + pV \tag{6-7}$$

Differentiate both sides and substitute the differential form of Helmholtz free energy.

$$dG = -SdT + Vdp \tag{6-8}$$

$$S(T,p) = \left(\frac{\partial G}{\partial T}\right)_p \tag{6-9}$$

$$V(T,p) = \left(\frac{\partial G}{\partial p}\right)_T \tag{6-10}$$

For a thermodynamic system, if the number of particles in the system will change, the newly added or reduced number of particles will also cause the system's internal energy to change. This change in internal energy can be expressed by formula (6-11).

If the change in the number of particles is taken into account, the differential representation of internal energy only needs to add a term due to the change in the number of particles behind.

$$dU = TdS - pdV + \mu dN \tag{6-11}$$

This way we can also get a very useful relation

$$\mu = \frac{G}{N} \tag{6-12}$$

From a microscopic point of view, we can also calculate the microscopic expression for the chemical potential (Pathria & Beale, 2011):

$$\mu(N,V,T) = k_B Tln\left[\frac{N_n}{V}\left(\frac{h^2}{2\pi m k_B T}\right)^{\frac{3}{2}}\right] \tag{6-13}$$

In the formula, k_B is Boltzmann constant, h is Planck constant, m is water molecular mass. The other N_n, V, T are molecular number, volume and temperature, respectively.

Of course this is for the microscopic form, which we can convert into the macroscopic sociothermodynamic form, that is

$$\mu(N,V,T) = \beta Tln\left[\frac{Nn_p}{V}\left(\frac{h^2}{2\pi m k_B T}\right)^{\frac{3}{2}}\right] \tag{6-14}$$

Example 6-1:

For a city with a population of 1 million and an area of 1,000 square kilometers, the social temperature is 272K. Then calculate its chemical potential.

Solution:

Substitute various parameters into the formula (6-14) for the chemical potential in this way:

$$\mu(N,V,T)=2\times10^{27}\times1.38\times10^{-23}\times272ln$$
$$\left[\frac{10^{6}\times2\times10^{27}}{3\times10^{9}}\left(\frac{6.63^{2}\times10^{-68}}{6.28\times3\times10^{-26}\times1.38\times10^{-23}\times272}\right)^{\frac{3}{2}}\right] \quad (6\text{-}15)$$

This allows us to calculate the chemical potential of the city as:

$$\mu(N,V,T)\approx-1.38\times10^{8}(J) \quad (6\text{-}16)$$

Math further

With the differential form of the change in the internal energy due to the change in the number of particles, we can also add the change in the number of particles to some other important thermodynamic variables. In this way, the Helmholtz free energy and Gibbs free energy in the thermodynamic system in which the number of particles or the number of social individuals changes can be expressed as

$$dH \doteq TdS+Vdp+\mu dN \quad (6\text{-}17)$$

$$dF=-SdT-pdV+\mu dN \quad (6\text{-}18)$$

$$dG=-SdT+Vdp+\mu dN \quad (6\text{-}19)$$

For the convenience of comparison, the definitions of enthalpy, Helmholtz free energy and Gibbs free energy are also written here.

$$H = U + pV \tag{5-61}$$

$$F = U - TS \tag{5-136}$$

$$G = F + pV \tag{5-138}$$

With the differential expression of these free energies, we can also list various calculation formulas of chemical potential according to the definition of partial differential.

$$\mu = \left(\frac{\partial U}{\partial N}\right)_{S,V} = \left(\frac{\partial H}{\partial N}\right)_{S,p} = \left(\frac{\partial F}{\partial N}\right)_{T,V} = \left(\frac{\partial G}{\partial N}\right)_{T,p} \tag{6-20}$$

Example 6-2:

If the chemical potential of a city $\mu(N,V,T) = -1.38 \times 10^8 J$, Under the condition of keeping entropy and social space constant, how much will the internal energy change for every additional 10,000 people?

Solution:
According to the definition of chemical potential, we can get:

$$\mu = \left(\frac{\partial U}{\partial N}\right)_{S,V}$$

therefore:

$$\Delta U = \mu \Delta N = -1.38 \times 10^8 \times 10^4 = -1.38 \times 10^{12} (J)$$

That is to say, for every additional 10,000 people, the city's internal energy will decrease by $-1.28 \times 10^{12} J$.
The above results are obtained with the entropy unchanged. The decrease in internal energy is mainly due to the need to maintain the constant entropy under the condition of population increase.

Example 6-3:

For a city with a population of 1 million and an area of 1,000 square kilometers, the social temperature is 272*K*. Find the Helmholtz free energy and Gibbs free energy of the city.

Solution:

First of all, from the calculation results of Example 5-8, we can find that in order to maintain the operation of the city, the daily input of internal energy is

$$dU = dQ = 9.45 \times 10^{12} \, (J)$$

After adding physical internal energy, the total internal energy of the city is

$$U = U_p + dU = 2.25 \times 10^{13} \, (J)$$

Considering that the city's social temperature is

$$T = 272K$$

Similar to Example 5-11, the entropy of the city can be calculated as

$$S = 5.8 \times 10^{11} \left(J \cdot K^{-1}\right)$$

This allows us to calculate the city's Helmholtz free energy as

$$F = U - TS = 2.25 \times 10^{13} - 272 \times 5.8 \times 10^{11} = -1.35 \times 10^{14} \, (J)$$

According to the equation of state of ideal gas, we can also calculate

$$pV = N\beta T = 7.51 \times 10^{12} \, (J)$$

In this way, the Gibbs free energy can be obtained as

$$G = F + pV = -1.27 \times 10^{14} \, (J)$$

6.2 Conditions for thermal equilibrium

6.2.1 Criteria for entropy

One of the most important tasks in the equilibrium state of social thermodynamic is to solve when to reach thermal equilibrium and what are the conditions for reaching thermal equilibrium.

In fact, the judgment of heat balance is like finding the maximum and minimum values in mathematics. The whole judgment process is mainly to derive the relevant thermodynamic function. After the derivation, if the first derivative is 0, it means that there is an extreme value at this position, and through the positive and negative signs of the second derivative, it can be judged whether the extreme value is the minimum value or the maximum value.

Math further

So far, we have several functions, including internal energy, entropy, Helmholtz free energy, Gibbs free energy, and more.

The most widely used of these functions are entropy, Helmholtz free energy and Gibbs free energy.

We can get the Taylor series expansion of the entropy function first.

$$S = S_0 + \frac{1}{1!}\delta S + \frac{1}{2!}\delta^2 S + \frac{1}{3!}\delta^3 S + \dots \tag{6-21}$$

It can be assumed that the S-function is a function of multiple thermodynamic parameters, which are denoted by x_i.

Then the first-order derivative of the S-function can be expressed as

$$\sum_{i=1}^{n} \frac{\partial S}{\partial x_i} \tag{6-22}$$

The above Taylor series expansion term is

$$\delta S = \sum_{i=1}^{n} \frac{\partial S}{\partial x_i} \delta x_i \tag{6-23}$$

It can be obtained by analogy

$$\delta^2 S = \sum_{i,j=1}^{n} \frac{\partial^2 S}{\partial x_i \partial x_j} \delta x_i \delta x_j \tag{6-24}$$

In this way, to judge whether the system has reached an equilibrium state, we can judge by the following two conditions.

$$\delta S = 0 \tag{6-25}$$

$$\delta^2 S < 0 \tag{6-26}$$

This means that at the thermal equilibrium position, there is a maximum value of entropy. Therefore, this is a criterion for judging whether the thermodynamic system has reached the equilibrium state by the maximum value of entropy. This is because in equilibrium thermodynamics, if a thermodynamic system has no external energy input, it will eventually run to a state of maximum entropy and stabilize.

When the above conditions are met, the system reaches a state of equilibrium. At this time, the internal energy, social space and the number of social individuals will no longer change, so there is

$$\delta U = 0, \delta V = 0, \delta N = 0$$

Example 6-4.

Two cities that are relatively close, because the development of the city leads to the fusion of the two cities into a city with a larger population. If both cities have a population of 1 million and an area of 1,000 square kilometers, the social temperature is both 272K. Calculate the entropies of the two cities before the fusion. Then calculate the entropy of the city after fusion.

Solution:
The entropy for both cities is

$$S_1 = S_2 \approx 5.8 \times 10^{11} J \cdot K^{-1}$$

Therefore, before fusion, the total entropy of the two cities is

$$S = S_1 + S_2 = 1.16 \times 10^{12} J \cdot K^{-1}$$

If the two cities are integrated, the population becomes 2 million and the area becomes 2,000 square kilometers. Then there are

$$S = N\beta \left\{ \frac{3}{2} lnT + ln \frac{V}{Nn_p} + \frac{3}{2} \left[\frac{5}{3} + ln\left(\frac{2\pi m k_B}{h^2} \right) \right] \right\} \approx 1.16 \times 10^{12} J \cdot K^{-1}$$

That is to say, entropy has not changed. This is consistent with the actual situation. After all, although the two cities are nominally integrated, the number of social individuals in the two cities and the urban area have not changed.

Example 6-5.

If a city with a population of 1 million and an area of 1,000 square kilometers has a social temperature of 272*K* and adiabatic expansion to 1,100 square kilometers, find the change in entropy. If the constant pressure expands to 1,100 square kilometers, find the change in entropy.

Solution:

The change due to entropy is not very obvious. In order to see some of the slight changes, we improve the calculation accuracy of entropy, we can get

$$S \approx 5.783 \times 10^{11} J \cdot K^{-1}$$

Then we calculate the adiabatic coefficient from the heat capacity calculated in Example 5-7 and Example 5-9

$$\gamma = \frac{C_p}{C_V} = 2 \tag{5-101}$$

From formula (5-99), we can get

$$T_1V_1^{2-1} = T_2V_2^{2-1} \tag{5-108}$$

$$T_2 = \frac{T_1V_1}{V_2} = \frac{272\times1000}{1100} \approx 247K$$

Then we find the change in entropy after adiabatic expansion of the city

$$S \approx 5.770\times10^{11}\left(J\cdot K^{-1}\right)$$

It can be seen that the entropy decreases under the condition of adiabatic expansion. This is the negative work done by social pressure due to adiabatic expansion. This results in a reduction in internal energy, which in turn results in a reduction in entropy.
If it is equal pressure expansion, according to Charles' law, we can get

$$T_2 = \frac{V_2}{V_1}T_1 = \frac{1100}{1000}\times272 = 299(K)$$

then we can get:

$$S \approx 5.85\times10^{11}\left(J\cdot K^{-1}\right)$$

Therefore, under the condition of equal pressure expansion, although the pressure does negative work, the system absorbs more heat in the environment, causing the temperature to rise, resulting in an increase in entropy.

Example 6-6.

If one city has a population of 1 million, the other city has 10 million, and the urban area is the same 1,000 square kilometers, and neither city has external social pressure, it means that both cities are agglomeration caused by intra-population pressure. However, the interaction between individuals in a city with a population of 1 million is relatively weak, resulting in only 1/100 of the normal social interaction within it. The population of 1 million will all move into the city with a population of 10 million. Then find the change in entropy.

Solution:

First of all, for a city with a population of 1 million, since the internal pressure is 1/100 weaker, according to formula (5-11)

$$T_{100} = \frac{a_s N}{\beta V} = \frac{1.5\times10^7\times10^6}{2.76\times10^4\times3\times10^9}\approx0.18(K)$$

And the social temperature of a city with a population of 10 million will reach:

$$T_{1000} = \frac{a_s N}{\beta V} = \frac{1.5\times10^9\times10^7}{2.76\times10^4\times3\times10^9}\approx180(K)$$

So the entropy of a city with a population of 1 million is

$$S_{100} = 10^6\times2.76\times10^4\times\left\{\frac{3}{2}ln0.18+ln\frac{3\times10^9}{10^6\times2\times10^{27}}+\frac{3}{2}\times\left[\frac{5}{3}+ln\left(\frac{2\pi mk_B}{h^2}\right)\right]\right\}$$
$$\approx2.74\times10^{11}\left(J\cdot K^{-1}\right)$$

The entropy for a city of 10 million is:

$$S_{1000} = 10^7\times2.76\times10^4\times\left\{\frac{3}{2}ln180+ln\frac{3\times10^9}{10^7\times2\times10^{27}}+\frac{3}{2}\times\left[\frac{5}{3}+ln\left(\frac{2\pi mk_B}{h^2}\right)\right]\right\}$$
$$\approx4.97\times10^{12}\left(J\cdot K^{-1}\right)$$

So the total entropy of the two cities is

$$S=S_1+S_2\approx5.25\times10^{12}\left(J\cdot K^{-1}\right)$$

If 1 million people all move into a city of 10 million people, the temperature in the city is

$$T_{1100} = \frac{a_s N}{\beta V} = \frac{1.5\times10^9\times1.1\times10^7}{2.76\times10^4\times3\times10^9} \approx 199(K)$$

Its entropy is:

$$S = 1.1\times10^7\times2.76\times10^4 \times\left\{\frac{3}{2}ln199 + ln\frac{3\times10^9}{1.1\times10^7\times2\times10^{27}} + \frac{3}{2}\times\left[\frac{5}{3} + ln\left(\frac{2\pi mk_B}{h^2}\right)\right]\right\}$$

which is

$$S \approx 5.49\times10^{12}\left(J\cdot K^{-1}\right)$$

It can be seen that the entropy increases after fusion. This also reflects the fact that, for a large-scale city, if a small number of migrants are to be integrated into the city, the interaction between individuals of the migrant population should be as small as possible. In this way, if the interaction between foreign individuals is not strong enough, the entry of these foreign populations will lead to an increase in the entropy of the entire society, making it easier to achieve a balanced state. From a cultural point of view, it is easier for the whole city to maintain the original urban culture. Of course, this also reflects the population agglomeration effect of a city. That is, with the entry of a large number of foreign social individuals without strong interaction with each other, the expansion of the urban area will lead to the continuous increase of the entropy of the entire city, and the social thermodynamic system will become more stable.

If the interaction between foreign population individuals is relatively strong, after entering the city, the entropy of the entire city will decrease instead. This means that the foreign population will not easily integrate into the city's culture.

There are examples of this in real life. For example, the Roma people in Europe have a history of more than a thousand years, but it seems that the Roma people have not yet fully integrated into the mainstream European culture (Cahn, & Guild, 2010). But when millions of Ukrainian refugees enter Europe in 2022, they appear to be somewhat welcomed. The reason for this may be that the internal social interaction of Roma is stronger, while Ukraine belongs to a social-oriented society, and its internal social interaction is weaker than that of the family-oriented society. From the calculation results of this example, Ukrainians should be easier to integrate into mainstream European society.

6.2.2 Free energy criterion

We have already introduced two free energies earlier. Therefore, when judging whether the conditions of thermal equilibrium are reached, we can also use the derivation of free energy to judge. When finding the free energy derivative, we can also judge whether the first derivative of the free energy is equal to 0. However, since the sign of free energy and entropy is exactly opposite, the criterion for judging the second derivative is just opposite to that of entropy.

For Helmholtz free energy, whether the equilibrium condition can be reached depends on whether the first-order derivative is 0 and whether the second-order derivative is greater than 0. That is

$$\delta F = 0 \tag{6-27}$$

$$\delta^2 F > 0 \tag{6-28}$$

Math further

The meaning of δ in the formula is the same as the change of entropy. which is

$$\delta F = \sum_{i=1}^{n} \frac{\partial F}{\partial x_i} \delta x_i \tag{6-29}$$

$$\delta^2 F = \sum_{i,j=1}^{n} \frac{\partial^2 F}{\partial x_i \partial x_j} \delta x_i \delta x_j \tag{6-30}$$

Under the condition of equilibrium, the main thing is that the temperature does not change, the volume does not change, and the number of particles does not change.

$$\delta T = 0, \delta V = 0, \delta N = 0$$

Similar conditions can apply to Gibbs free energy.

$$\delta G = 0 \tag{6-31}$$

$$\delta^2 G > 0 \tag{6-32}$$

The judgment result of the Gibbs free energy is different from the judgment result of Helmholtz free energy. When use Gibbs free energy to determine whether the thermal balance is reached, the main thing is that the temperature of the thermodynamic system will not change, the pressure will not change, and the number of particles will not change.

$$\delta T = 0, \delta p = 0, \delta N = 0$$

It can be seen that the objects targeted by the three judgment methods are different, that is to say, different thermodynamic equilibrium conditions are achieved. This is why we need some other form of free energy function besides entropy. In this way, we can study different social thermodynamic systems through the combined judgment of multiple functions.

Example 6-7.

According to the conditions of Example 6-6, calculate the change of Helmholtz free energy before and after urban integration. The change of internal energy in the city can refer to the conditions and solution methods of Example 5-8. A city with a population of 1 million has an annual power generation capacity of 5 billion kWh, and a city with a population of 10 million has an annual power generation capacity of 50 billion kWh. The average person consumes 1,500 calories per day.

Solution:
From Example 5-8, the total internal energy of a city with a population of 1 million and 1,000 square kilometers is:

$$U_{100} \approx 2.25 \times 10^{13} (J)$$

In a city with a population of 10 million, the internal energy of the human thermodynamic system is:

$$U_p = \frac{3}{2} N\beta T = 1.5 \times 10^7 \times 2.76 \times 10^4 \times 310 \approx 1.3 \times 10^{14} (J)$$

$$dU = dE_{nutrition} + 50\% \times dE_{nutrition} \approx 9.45 \times 10^{13} (J)$$

$$U_{1000} = U_p + dU \approx 2.25 \times 10^{14} (J)$$

Solved by the formula for Helmholtz free energy:

$$F = U - TS$$

can get

$$F_{100} = U_{100} - T_{100}S_{100} = 2.25 \times 10^{13} - 0.18 \times 2.74 \times 10^{11} \approx 2.25 \times 10^{13}\,(J)$$

$$F_{1000} = U_{1000} - T_{1000}S_{1000} = 2.25 \times 10^{14} - 180 \times 4.97 \times 10^{12} \approx -6.7 \times 10^{14}\,(J)$$

therefore

$$F = F_{100} + F_{1000} = -6.48 \times 10^{14}\,(J)$$

If a population of 1 million is integrated into a city of 10 million, then:

$$U_{1100} = 2.48 \times 10^{14}\,(J)$$

$$F = F_{1100} = U_{1100} - T_{1100}S_{1100} = 2.48 \times 10^{14} - 199 \times 5.49 \times 10^{12} \approx -8.44 \times 10^{14}\,(J)$$

Therefore, the Helmholtz free energy is reduced, which indicates that the social thermodynamic equilibrium has become more stable. Compared with the change of entropy, the change of Helmholtz free energy is more obvious. The solution for Gibbs free energy is similar.

6.2.3 Case study: The impact of industrialization on a small town

In Asia, a very traditional small town can be thousands of years old. In the long historical development process, this ancient town still maintains a very traditional culture, which can even extend to the ancient Hemudu culture more than 7,000 years ago. In this small town, various farming tools from the Iron Age are still preserved 100 years ago. In addition, when building houses, ancient architectural customs and methods are still maintained.

In such a self-sufficient farming society, all energy use comes from plant nutrition. Including all kinds of vegetables grown by themselves as well as livestock, animals and so on. And heating, burning, cooking, etc. are derived from plants. So this farming society actually gets all its energy from solar energy. Its energy consumption is very low. The whole ecological cycle is also very good. This is why such a society can last for thousands of years.

However, with the advent of the global industrialized society, the large-scale application of various power equipment and internal combustion engine equipment has made a great change in the way of energy acquisition in this farming society. In this change, a large amount of electric energy is mainly delivered by equipment such as coal-fired and hydroelectric power generation. In addition, tools such as vehicles and more electrical equipment were introduced. As a result, the energy consumption of this farming society has increased rapidly. Of course, with the rapid increase in these industrial facilities, people's nutrition has also improved. This also means that the nutrition obtained by the social individuals in this small town has also reached a higher level.

This case analyzes the impact of industrialization on this town by calculating various thermodynamic parameters, including changes in internal energy, free energy and entropy, in a town with a strong traditional culture.

This small town is located at 28 degrees north latitude, with a population of 20,000 and an area of 70 square kilometers.

Before industrialization, the town had no power plants, so the main sources of energy input were food and burning plants. Considering that because the food source is relatively simple, the physiological calorie consumption of the human being is lower than that of the individual in the industrialized society.

According to Example 5-8, we first calculate the basic internal energy of the human physiological system. According to the calculation that the human body is entirely composed of water molecules and the average physical temperature of the human body is 310*K*, then:

$$U_p = \frac{3}{2} N\beta T = 1.5 \times 20000 \times 2.76 \times 10^4 \times 310 \approx 2.57 \times 10^{11} (J)$$

Then consider that this is a traditional farming society, so the nutrition intake of each person is less than that of an industrialized society. Here, based on an average of 1,200 kcal per person, the nutritional energy required to maintain the operation of a traditional society is:

$$dE_{nutrition} = 1200 \times 4184 \times 20000 \approx 1.0 \times 10^{11} (J)$$

Burning plants for energy can be calculated according to the energy required to boil 20 liters of water with an average temperature of 20 degrees Celsius per person per day

$$dE_{burn} = 4.18 \times 80 \times 20 \times 1000 \times 20000 \approx 1.34 \times 10^{11} (J)$$

Of course, the energy produced by burning plants is not used efficiently by the human body. The energy consumed by burning plants mainly meets the needs of the human body for functional movement, the needs of ambient temperature, and the ability to obtain nutrients from food more efficiently. Among these energy requirements, nutritional requirements have been reflected in the energy items of nutritional gain. The energy consumed by the body's exercise accounts for about 50% of the basal metabolic needs. More importantly, the energy obtained by these burning plants is not used to drive other mechanical power devices, especially transportation. Therefore, we can find the daily internal energy change of this traditional society:

$$dU = dE_{nutrition} + 0.5dE_{nutrition} \approx 1.5 \times 10^{11} (J)$$

In this way, the total energy consumed by the small town can be calculated as:

$$U = U_p + dU \approx 4.07 \times 10^{11} (J)$$

In addition, we can also calculate the heat Capacity at constant volume of this traditional society:

$$C_v = N\beta = 20000 \times 2.76 \times 10^4 = 5.52 \times 10^8 \left(J \cdot K^{-1}\right)$$

Then according to the formula

$$dU = C_V dT$$

The change in temperature caused by the change in internal energy can be calculated as:

$$dT = \frac{dU}{C_v} = \frac{1.5 \times 10^{11}}{5.52 \times 10^{8}} \approx 272(K)$$

That is to say, the input of these energy can maintain the society at a social temperature of 272*K*. This is not very different from the physical temperature of the environment, indicating that this society and the surrounding environment have reached a certain balance of energy supply. In fact, this also means that if the average nutrient level of the villagers increases, it may easily cause the change of social temperature to exceed the ambient temperature, which will make it more difficult for the society to obtain more energy from the environment.

From this we can also see that in the traditional society, after thousands of years of development, a certain balance has been reached between human and nature. The efficiency with which humans obtain energy from nature has reached a very high level. Therefore, if there are some changes in this society, causing this balance to be broken, various problems may arise.

For example, the collective farm system implemented in the Soviet Union and Southeast Asia in the middle of the 20th century severely damaged the enthusiasm of farmers to produce food. Without the enthusiasm to produce food, it means that the ability of the whole society to obtain energy from the environment declines rapidly. The original balance was broken, and the new mechanism could not provide more sufficient energy, which caused a large number of famines. For example, because the Soviet Union implemented the collective farm system, the Great Famine in Ukraine in 1932–1933 caused more than 3 million people to starve to death and caused a large number of malnutrition problems for babies born that year (Werth, 2008; Lumey, Khalanggot & Vaiserman, 2015).

Then we can also find the entropy, Helmholtz free energy and Gibbs free energy of this traditional society. These free energies are important indicators for judging thermal balance.

The first is the entropy of this society:

$$S = \frac{3}{2}N\beta lnT + N\beta ln\frac{V}{Nn_p} + \frac{3}{2}N\beta\left[\frac{5}{3} + ln\left(\frac{2\pi mk_B}{h^2}\right)\right] \quad (5\text{-}121)$$

Substituting the above data, we can get:

$$S \approx 1.23\times 10^{10}\left(J\cdot K^{-1}\right)$$

We can calculate the Helmholtz free energy of the small town as:

$$F = U - TS = 4.07\times 10^{11} - 272\times 1.23\times 10^{10} \approx -2.93\times 10^{12}\,(J)$$

According to the Van der Waals equation (5-9), it can be calculated:

$$p = \frac{2.76\times 10^4\times 20000}{V - 2000}T - \frac{1.5\times 10^9\times 20000^2}{V^2} \approx 701(Pa)$$

$$pV = 701\times 3\times 70\times 10^6 = 1.47\times 10^{11}\,(J)$$

In this way, the Gibbs free energy can be obtained as:

$$G = F + pV = -2.78\times 10^{12}\,(J)$$

Let's analyze it again, if industrial technology is gradually applied to this small town. Mainly reflected in the following aspects:

1. A power plant was built, which increased the energy supply of the small town rapidly. The power station uses water power, wind power, solar power and fuel oil to generate 100 million kWh of electricity. After using power plants, the efficiency of energy utilization is much higher than that of burning plants, which also means that in this traditional society, more energy can be obtained from the environment. For example, the power of water

in the past was not absorbed by this society in the past. But with the use of hydroelectric power, the energy has been reused.

2. Purchase a large number of vehicles, including trucks, buses and cars. The number of various vehicles reached 2,000.
3. A solid building with reinforced concrete structure was built. A wide concrete asphalt road was built. About one meter of road requires 1000 kilograms of material. The average person lives in a house that uses 10,000 kilograms of building materials.
4. More electrical equipment is used.

Among these devices, roads that have already been built cannot be moved, so it is only necessary to calculate the energy required for the construction process and the energy required for daily maintenance. These buildings are calculated at 200,000 kWh per capita and have a life span of 100 years. The average daily energy consumption is:

$$dE_{house} = \frac{2\times10^5 \times 3.6\times10^6}{100\times365} \times 2\times10^4 \approx 3.95\times10^{11}\,(J)$$

Cars can move, so they have kinetic energy. Calculated according to the average consumption of 2 liters of gasoline per car, the energy generated per liter of gasoline is $4.5\times10^7\,J$, the efficiency of the car engine is 50%, and the kinetic energy of these cars reaches

$$dE_{car} = 2000\times4.5\times10^7 \times 50\% \times 2 = 9\times10^{10}\,(J)$$

Because power plants can provide additional energy, and these energy will be consumed by vehicles, electrical equipment, and mechanical equipment. Although the efficiency of energy utilization is not necessarily high. It can be calculated according to the energy utilization rate of 50%. In this way, by converting the annual power generation to the daily power generation, it is possible to calculate the energy input of the whole small town increased due to the emergence of the power plant as,

$$dE_{electric} = \frac{1\times10^{8}\times3.6\times10^{6}}{365}\times50\% \approx 4.9\times10^{11}\,(J)$$

This energy is generated for the needs of various household appliances, lighting, factory production, etc.

We can note that only part of the energy that is expended is to provide the social individual with better access to nutrients. In addition, there is energy such as automobiles, which partly provide the kinetic energy of social individuals. In this way, the kinetic energy of the car and the use of various electrical appliances directly contribute to the social thermodynamic parameters, which improves the energy efficiency of social individuals.

According to the ratio of passengers and car mass, it can be estimated that about 15% of the kinetic energy of the car can be converted into the kinetic energy of social individuals. The kinetic energy of this part is:

$$dE_{passenger} = 15\%\times dE_{car} \approx 1.35\times10^{10}\,(J)$$

After the use of various electrical equipment, the efficiency of human acquisition of nutrients has increased, and the daily intake of calories can reach 1,500 kcal, then:

$$dE_{nutrition} = 1500\times4184\times20000 \approx 1.26\times10^{11}\,(J)$$

Of course, another effect of industrialization is that social individuals have a greater ability to absorb energy other than nutrition. Therefore, the energy consumption of this part may be far more than 0.5 times the energy required for nutrition. However, for the convenience of calculation, 0.5 times the energy consumption of nutrition is still used. In this way, the internal energy change of the whole society will increase to:

$$dU = dE_{nutrition} + 0.5dE_{nutrition} + dE_{passenger} \approx 2.03\times10^{11}\,(J)$$

However, if the residents of the small town do not change their production and lifestyle, the overall social temperature will increase:

$$dT = \frac{dU}{C_v} = \frac{2.03 \times 10^{11}}{5.52 \times 10^{8}} \approx 367(K)$$

This temperature has exceeded the ambient temperature. This also means that it becomes more difficult for this society to obtain energy from the environment in accordance with the requirements of thermodynamic equilibrium. Traditional farming methods cannot continue. This is also the reason why many traditional farmers had to give up farming after industrialization. Because the cost of living has risen rapidly due to industrialization, the benefits obtained from traditional farming methods will not be able to maintain the daily living expenses of farmers. The only way is to enter the industrial chain and become a member of it, so as to maintain the basic balance of payments for a family.

In the above calculation process, in addition to meeting the nutritional and kinetic energy needs of social individuals, in fact, cars, houses, electrical appliances, etc. consume a lot of energy. These energies are:

$$dE_e = dE_{house} + dE_{car} + dE_{electric} - dU \approx 7.75 \times 10^{11} (J)$$

This energy actually turns into heat and is released into the air, causing the atmospheric temperature to rise.

Specifically, we can estimate it on a daily basis. Because for the duration of the day, the air above the town can basically be seen as an adiabatic system. In this adiabatic system, the energy generated is converted into heat in the air. However, we can also notice that some of these energies are hydro and wind power, which are all green energy. According to the current situation, green energy accounts for about 20% of the total energy. So this energy is released into the air, 80% of which will increase the temperature of the atmosphere.

It is now known that the total mass of the earth's atmosphere is about $5.14 \times 10^{18} kg$ (Verniani, 1966; Trenberth, 1981, 1994), and the surface area of the earth is 5.1×10^{8} square kilometers. This way we can calculate the mass of the air over the town as

$$m \approx \frac{70}{5.1 \times 10^{8}} \times 5.14 \times 10^{18} = 7.06 \times 10^{11} (kg)$$

Calculate according to the specific heat capacity of air $1006 J \bullet kg^{-1} K^{-1}$. This amount of energy can be calculated to cause the town's atmospheric temperature to rise over the course of the day

$$dT = \frac{dE_e}{C_v} = \frac{7.75 \times 10^{11} \times 80\%}{1006 \times 7.06 \times 10^{11}} \approx 0.00087 (K)$$

The effect of this temperature increase does not seem to be very obvious, but if it is accumulated for 365 days in a year, it can be found that this temperature will have a very large impact on the atmospheric temperature. On average, the temperature of the atmosphere increases by about 0.3 degrees per year. Of course, more than 70% of the Earth's surface is ocean. There are not so many densely populated cities. Current data on global temperature increases suggest that global temperatures have risen by more than one degree in 100 years (Burns, 1997).

Of course these additional energy inputs are necessary. Because we can see from the effects of the heat engine, if the heat or energy is to be transferred from the low temperature system to the high temperature system, work must be done to consume additional energy, so that the cooling or heating effect can be achieved.

Of course, in an industrialized society there is also an agricultural foundation. Instead, modern farming techniques have used mechanical farming techniques to achieve higher grain yields.

We can also calculate the entropy and free energy after industrialization.

Due to the rise in social temperature, entropy becomes:

$$S_1 = 20000 \times 2.76 \times 10^4 \times \left\{ \frac{3}{2} ln367 + ln \frac{3 \times 70 \times 10^6}{20000 \times 2 \times 10^{27}} + \frac{3}{2} \left[\frac{5}{3} + ln \left(\frac{2\pi m k_B}{h^2} \right) \right] \right\}$$
$$\approx 1.25 \times 10^{10} \left(J \cdot K^{-1} \right)$$

In addition, the heat released into the air also includes entropy, which can be calculated according to the formula of entropy. The atmospheric temperature is taken as 300K.

$$dS_2 = \frac{dE_e}{T} = \frac{7.75 \times 10^{11}}{300} = 2.58 \times 10^9 \left(J \cdot K^{-1} \right)$$

In this way, we can calculate the entropy after industrialization as

$$S = S_1 + dS_2 \approx 1.51 \times 10^{10} \left(J \cdot K^{-1} \right)$$

It can be seen that the entropy has increased after industrialization. This meets the requirements of thermodynamic balancing.

After industrialization, internal energy became

$$U = U_p + dU = 4.6 \times 10^{11} (J)$$

Helmholtz free energy is

$$F = U - TS = 4.6 \times 10^{11} - 367 \times 1.51 \times 10^{10} \approx -5.08 \times 10^{12} (J)$$

According to the Van der Waals equation (5-9), it can be calculated:

$$p = \frac{2.76 \times 10^4 \times 20000}{V - 2000} T - \frac{1.5 \times 10^9 \times 20000^2}{V^2} \approx 951(Pa)$$

$$pV = 2772 \times 70 \times 10^6 = 1.94 \times 10^{11} (J)$$

In this way, the Gibbs free energy can be obtained as:

$$G = F + pV = -4.88 \times 10^{12}\,(J)$$

It can be seen that both Helmholtz free energy and Gibbs free energy have decreased. This is also an important sign that the society has regained thermodynamic equilibrium after industrialization.

Of course, in order to reduce the social temperature, the heat capacity can be increased. From the calculation Formula (5-70) of heat capacity, it can be seen that as long as the number of people N is increased, the heat capacity can be rapidly increased, thereby reducing the social temperature. And increasing the population on the same area of land means that the scale of the small town is expanding. After expanding to a certain extent, a city is formed. The agglomeration of population in a city can significantly increase the thermal capacity of the entire city, and when the same energy is consumed, the social temperature will not increase significantly.

In addition to increasing the population N, to reduce the social temperature, we can also develop in the direction of reducing the social space. For example, the population is gradually concentrated to the location of the core urban area, so that the social space of people's life will be greatly reduced. And those farmlands, etc., which were previously used for farming, as well as mountains, etc., can be used for special cultivation of food and trees, rather than providing space for human activities.

In this way, we can see from the Van der Waals equation that as the social space decreases, the temperature also decreases in the same proportion. This is also an important way for the gradual urbanization of many rural areas.

6.3 Phase equilibrium

6.3.1 Phase diagram

A phase diagram is a graphical representation of how the state of matter changes with changes in pressure and temperature. Therefore, the phase diagram mainly includes two coordinates of temperature (Kelvin) and pressure. In the coordinates of the phase diagram, the whole diagram is divided into several blocks with the phase equilibrium line as the dividing standard. Each block will represent a state of matter. such as solid, liquid and gaseous.

Figure 6-1 shows the phase diagram of a substance.

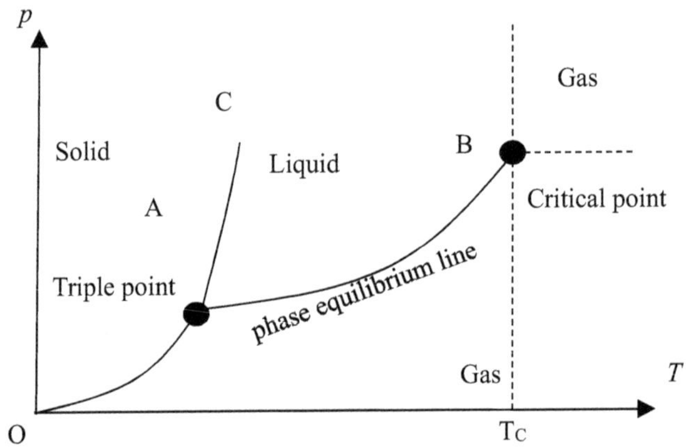

Figure 6.1. A typical phase diagram

It can be seen from Figure 6-1 that the substance has three phases, namely gas phase, liquid phase and solid phase. It can be seen from the figure that when the pressure is relatively small, even if the temperature is very low, the substance is in the state of gas. But if the pressure is increased when the matter is cooler, this can make the matter solid. From the figure, we can also find a very interesting phenomenon, that is, under the condition that the temperature and pressure are relatively low, it can directly change from a solid state to a gas state, and of course it can change from a gas state back to a solid state. This substance is also relatively common in daily life. For example, the solid carbon dioxide used in fire extinguishers directly converts carbon dioxide into solid by increasing the pressure. When the fire is extinguishing, the pressure valve is fully opened, and the pressure of carbon dioxide can be reduced at this time, so that the solid carbon dioxide can be rapidly converted into gaseous carbon dioxide to achieve the purpose of fire extinguishing. It can also be seen from the phase diagram that under moderate pressure and temperature, the substance can also become a liquid state.

In the phase diagram, the curve dividing the different phases of the substance in the figure is the position where the two phases are in equilibrium, such as the AB or AC curve in the figure. The intersection point A of the three curves in the middle is called the three-phase equilibrium point, or triple point. At a two-phase or three-phase equilibrium point, it is usually a situation where two or three states of matter exist at the same time. For example, when water is boiling, it is such a two-phase equilibrium position. When water boils, both water vapor and liquid water exist. At this time, the temperature of the water vapor and the liquid water are also the same. That is, under the condition of one atmospheric pressure, the temperature of boiling water and water vapor are both 100 degrees Celsius.

6.3.2 Three social states

Because matter exists in three states: gaseous, liquid and solid. Since the social system is also a thermodynamic system, theoretically, the social thermodynamic system should also have three states, namely gaseous society, liquid society and solid society.

The p-V curve of a gaseous society is similar to the p-V curve of an ideal gas, and can be approximately described by the equation of state of an ideal gas. Due to the instability of the gas, it is very sensitive to changes in various thermodynamic parameters. As long as the temperature, pressure and social space in it change, other thermodynamic parameters of this gaseous society can easily change with it. In addition, the fluidity of the gaseous society is also relatively strong. As long as there is a thermodynamic imbalance, it can flow rapidly with changes in pressure and temperature gradients. Unlike the other two states of society, the gaseous society also has the ability to rapidly expand and contract. That is, under the condition of great external pressure, the social space of the gaseous society will be rapidly compressed. When the internal pressure is relatively high, the gaseous society is also easy to expand.

Liquid society is a type of condensed state. Due to the strong interaction within the society, the members of the society are connected more closely together, forming a relatively solid overall structure. Liquid society also has relatively strong mobility. But unlike the gaseous society, the liquid society can always maintain the integrity among the members of the society during the flow process. Although the shape of the liquid society is relatively easy to change, the compression and expansion coefficients of the liquid society are relatively small. Even if the external or internal pressure changes, the liquid society is not easy to expand or contract. Therefore, compared with the gaseous society, the liquid society is a more stable social state.

Solid society means that all social members are fixed in a specific social space. Like liquid societies, solid societies cannot easily expand or compress. But unlike liquid societies, solid societies are basically immobile. Although solid society is a very stable social structure, the interaction between individual members in solid society is limited to other basic society members in adjacent social spaces.

All three social forms can be described by the phase diagram in Figure 6-1. As can be seen from Figure 6-1, the three social states are different from the three different types of social types such as family type, social type and balance type. The three social states can be transformed into each other. This is mainly affected by social temperature and social pressure. If the social temperature is relatively low and the social pressure is relatively high, it is possible for a gaseous society to transform into a liquid society. vice versa. The three different types of society are mainly related to the form of interaction between members of the society. And this form of interaction includes both social interaction and gender interaction. The differences in these interactions are mainly influenced by different social and cultural backgrounds. In order to form a unique cultural background of a society, it takes a long process of cultural inheritance. Therefore, the three different types of society are more difficult to change. Naturally there are also great difficulties in changing from one type of society to another.

All three social states can be described by the Van der Waals equation. Therefore, we can obtain some important conditions for the mutual transformation of the three social states by analyzing the Van der Waals equation.

6.3.3 Critical temperature

We can slightly modify the Van der Waals equation to adapt it to the macroscopic social thermodynamic system.

$$\left(p+\frac{a_s N^2}{V^2}\right)(V-b)=N\beta T \tag{5-9}$$

Where a_s reflects the interaction between social individuals. N is the number of individuals in the whole society. V is the social space of the social thermodynamic system, p is the pressure of the system, b is the social space occupied by all individuals in the social thermodynamic system, and β is the molecular number of water contained in the social individual multiplied by the Boltzmann constant. T is temperature.

Assuming the temperature is constant, since this is a nonlinear equation, we take the derivative of the volume, which allows us to determine some important inflection points of the curve to accurately draw the curve. The first derivative of the inflection point must be 0, so:

$$\frac{(V-b)dp}{dV}=p-\frac{a_s N^2}{V^2}+\frac{2a_s N^2 b}{V^3}=0 \tag{6-33}$$

$$p=\frac{a_s N^2}{V^2}-\frac{2a_s N^2 b}{V^3} \tag{6-34}$$

This is the pV curve corresponding to the inflection point of the curve for different temperatures. It can be seen that this is a parabola, corresponding to points D and E at different temperatures. And the curves of points B and C at different temperatures should also be similar. Therefore, the dashed parabolas corresponding to points B and C at different temperatures are drawn in Figure 6-2.

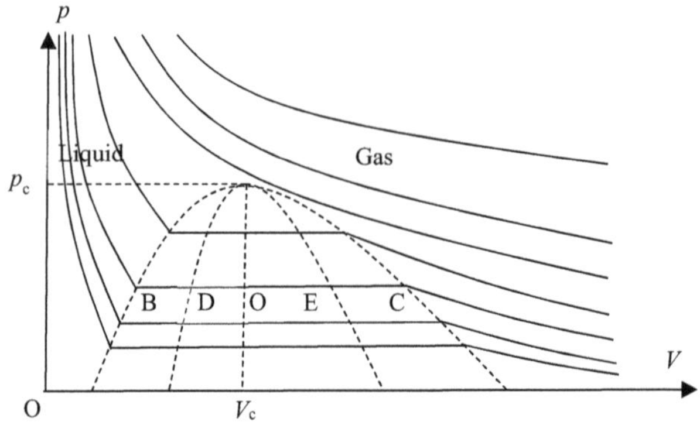

Figure 6.2. Isotherm curves

Different curves appear due to temperature changes. Some have inflection points, and some have no inflection points. The existence of these inflection points should be related to the first derivative of formula (6-34). Therefore, we take a derivative of V with respect to formula (6-34) and set it to 0, we can get:

$$\frac{dp}{dV} = -\frac{2a_s N^2}{V^3} + \frac{6a_s N^2 b}{V^4} = 0 \tag{6-35}$$

This gives:

$$V_C = 3b \tag{6-36}$$

We can call this volume the critical volume. This volume reflects the volume of the Van der Waals curve at the point of inflection.

Substituting the critical volume into equation (6-34), we can obtain the critical pressure.

$$p_C = \frac{a_s N^2}{9b^2} - \frac{2a_s N^2 b}{27b^3} = \frac{a_s N^2}{27b^2} \tag{6-37}$$

The critical temperature can then be found by substituting the critical pressure and critical volume into the Van der Waals equation (5-9):

$$T_C = \frac{8a_s}{27\beta v_p} \tag{6-38}$$

where v_p is the size of the social space occupied by a single social individual. And

$$v_p = \frac{b}{N}$$

From the calculation formula of critical temperature (6-38), it can be seen that the critical temperature of social thermodynamic system is mainly related to three factors. The stronger the interaction among social individuals, the higher the critical temperature, and vice versa. The larger the social space occupied by a single social individual, the lower the critical temperature, and vice versa. The higher the number of water molecules contained in a single social individual, the lower the critical temperature, and vice versa.

When the temperature of the thermodynamic system is less than the critical temperature, it means that the curve has two inflection points. That is, there is a maximum value and a minimum value. The curve is similar to the graph shown in Figure 1-3. And if the temperature of the system is greater than the critical temperature, it means that the curve has no inflection point.

Example 6-8.

For a society with a population of 1 million and 1000 square kilometers, find the critical pressure, social space and social temperature of the society according to the Van der Waals equation.

Solution:

According to the volume occupied by each social individual of 0.06 cubic meters, first find the social space occupied by the social individual:

$$b = Nv_p = 10^6 \times 0.06 = 6 \times 10^4 m^3$$

This gives the critical volume:

$$V_c = 3b = 1.8 \times 10^5 m^3$$

Recombination parameters

$$a_s = 1.5 \times 10^9 Pa \cdot m^6$$

The critical pressure can be found:

$$p_C = \frac{a_s N^2}{27b^2} = \frac{1.5 \times 10^9 \times 10^{12}}{27 \times 10^{12} \times 0.06^2} = 1.54 \times 10^{10} Pa$$

set the parameter

$$\beta = n_p k_B = 2.76 \times 10^4 J \cdot K^{-1}$$

Substituting into formula (6-38), we can find the critical temperature as

$$T_C = \frac{8a_s}{27\beta v_p} = \frac{8 \times 1.5 \times 10^9}{27 \times 2.76 \times 10^4 \times 0.06} = 2.68 \times 10^5 (K)$$

It can be seen that this is a very high critical temperature. Considering that the temperature of a social thermodynamic system is generally lower than the critical temperature, as long as the social space is not too large, the society is basically in a condensed state similar to solid or liquid. At the same time, under the premise of appropriate pressure and social space, there should be relatively large differences in social status.

6.3.4 Actual isotherms

The critical temperature obtained from the Van der Waals equation shows that when the temperature is less than the critical temperature, there will be two inflection points in the pressure-volume curve, the shape of which is shown in Figure 1-3. Now we divide the two inflection points with a dashed line BC, and make sure that the areas of the BDO and OEC shapes are equal, thus forming a dashed line. This equal-area segmentation method is called Maxwell's equal-area method.

The BC straight line is the isotherm of the actual thermodynamic system at the phase equilibrium position. The Van der Waals isotherm BDOEC is the result of theoretical calculation, which is unstable. We can use the free energy criterion to show that the BC isotherm is more stable than the BDOEC isotherm.

We can map the Van der Waals isotherm to the coordinates of free energy to form the variation of Helmholtz free energy as shown in Figure 6-3.

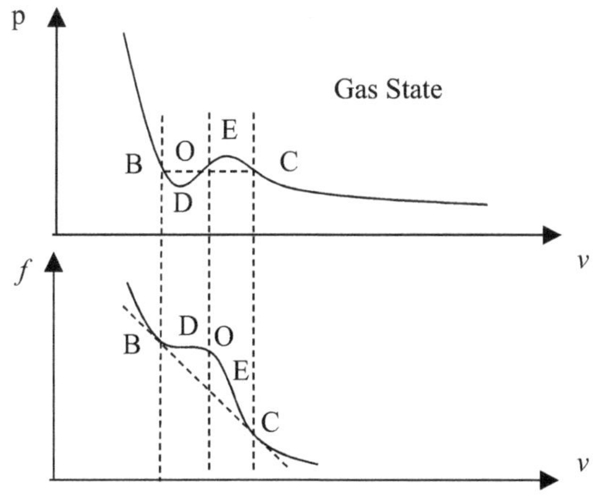

Figure 6.3. A typical solution of van der Waals equation

As can be seen from Figure 6-3, BDOEC corresponds to the following free energy f changes in the Van der Waals curve. If it is the BC dotted line in the isotherm of the actual thermodynamic system, it corresponds to the BC oblique line of the Helmholtz free energy below. If the Van der Waals isotherm changes from B to D to OEC, it corresponds to the curve BDOEC of the Helmholtz free energy below.

From the change of Helmholtz free energy, the change of free energy of BDOEC is larger than the change of Helmholtz free energy corresponding to the BC straight line. Therefore, according to the Helmholtz free energy criterion of the equilibrium state, the system is always in the state with the minimum free energy, which explain that the actual state change is located above the BC dotted line of the Van der Waals curve. At the position of the BC line, the pressure of the thermodynamic system does not change, but the volume changes significantly. This is the case with water in a boiling state. At the same time, we can also notice that the temperature also remained stable during this period. Therefore, in the interval B to C, the actual isotherm must be the BC straight line, not the BDOEC curve.

The BC isotherms correspond to the AB or AC curves in Phase Diagram 6-1. This is where the gas-liquid or solid-liquid two phases are in equilibrium.

Example 6-9.

If the cities in Example 6-8 generally use Web2.0 technology, the radius of individual social space expands, but the social interaction between individuals does not change, find the critical social space, pressure and temperature. Draw the Van der Waals isotherm with temperature T=27*K*, and draw the actual isotherm when two phases coexist.

Solution:

After using information technology, a person's social space will be expanded rapidly. According to the estimation in Section 4.3, if the Web2.0 technology is used, the radius of a person's social space can reach 10*m*. In this way, the parameter *b* in the Van der Waals equation needs to be modified, namely:

$$b = 1.7N\pi r^2 \approx 1.7 \times 10^6 \times \pi \times 10^2 \approx 5.34 \times 10^8 \left(m^3\right)$$

In this way, the critical volume can be obtained as:

$$V_c = 3b \approx 1.6 \times 10^9 m^3$$

$$p_C = \frac{a_s N^2}{27b^2} = \frac{1.5\times10^9\times10^{12}}{27\times5.34^2\times10^{16}} \approx 195Pa$$

$$T_C = \frac{8a_s}{27\beta v_p} = \frac{8\times1.5\times10^9}{27\times2.76\times10^4\times534} \approx 30(K)$$

It can be seen that the critical temperature is significantly reduced, that is to say, under this condition, the city must be able to enter the condensed state when the social temperature is very low. The limit of this temperature is 30*K*. If it cannot enter the condensed state, it can be regarded as an unstable factor in the city. After the use of web2.0, the social space of social individuals increases, which means that if the interaction between social individuals is not enhanced, the city will become unstable.

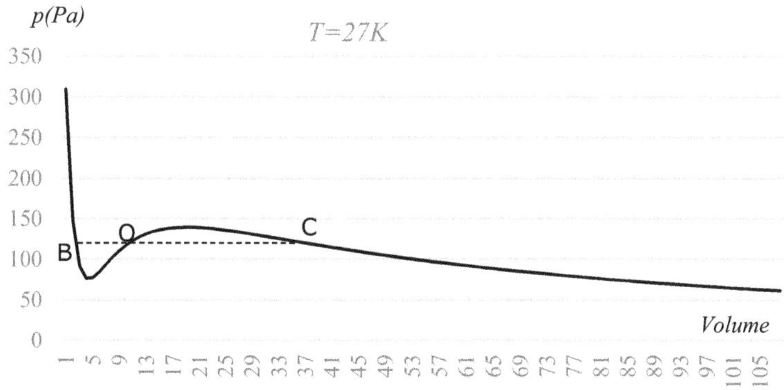

Figure 6.4. The van der Waals curve of a social system

In Figure 6-4, the volume conversion is $V=(7.5+0.9\cdot Volume)\times10^8(m^3)$. The pressure of the BOC isotherm is $p=110Pa$. The volume of point B is $V=1.02\times10^9m^3$. Point O is $V=1.65\times10^9m^3$. Point C is $V=4.17\times10^9m^3$.

As can be seen from Figure 6-4, after the use of information technology, under the condition of relatively low social temperature, the figure is T=27*K*, if the pressure is reduced to about 120*Pa*, the system will enter an isobaric state at this time. In this isobaric state, despite the large-scale changes in the social space of the city, the pressure of the whole society will always remain constant. If this social space change is caused by the application of information technology. It shows that in this state of equal pressure, the development of information technology will not cause changes in the pressure of the whole society.

But we can also notice that on the right side of the actual BC isotherm, that is, to the right of point C, if the pressure drops further and is lower than the position of the BC isotherm, a little change in pressure will lead to a wide range of social space. At this time, although social pressure is decreasing, the decrease is not obvious. This is similar to the gaseous state in the state of matter. If this kind of social space is also caused by the application of information technology, it means that the application of information technology will not cause a very obvious change in social pressure. Of course, it can also be seen from the pressure as an independent variable. If it is on the right side of point C, it means that a small change in pressure will cause a large-scale change in social space. This kind of social expansion and contraction ability will be relatively strong.

However, if it is on the left side of the BC line, that is, on the left side of point B, we can see that a small change in the social space on this side will lead to a large range of changes in the entire social pressure. This is similar to the condensed state in the state of matter. It shows that the development of information technology in this area will cause rapid changes in the pressure of the whole society, which will lead to obvious anomalies within the society. In the case of such a condensed state, for example, any small change in the way information is delivered can cause large-scale changes in social pressure.

Conversely, if pressure is used as an independent variable, the left side of point B means a wide range of changes in social pressure, and the impact on social space is not very large. Such a society has stronger stability and is less susceptible to external influences.

6.3.5 Equilibrium phase transition

Considering that the critical temperature obtained in Example 6-9 is relatively high, that is to say, the general social thermodynamic system is basically below the critical temperature. If a thermodynamic system is below a critical temperature, it means that it must have two phases.

First of all, it can be determined that the two inflection points should be on both sides near the critical volume. The lower the temperature, the greater the range of variation in this volume. Within this isotherm volume change range, neither the pressure nor the temperature of the thermodynamic system will change.

From the Van der Waals isotherm, two "phases" in the system can be clearly decomposed. The main difference between them is that the relationship between pressure and social space is different. One of the phases has a very steep curve. This means that with the large changes in pressure, the changes in social space are not large, and the ability of society to withstand pressure is relatively large. At this time, the interaction between social individuals is relatively large, and the integration between social individuals is relatively close. This is very consistent with the condensed state of matter. But we can also notice another very interesting phenomenon, that is, if the technology of the whole society develops, especially the development of information technology. It will lead to the increase of social space, and the effect of the increase of information space is consistent with the increase of physical space. If it is in this condensed state, it means that the increase in social space caused by information technology will cause rapid changes in social pressure. Naturally, it is easier to enter the flatter phase on the right side of the curve. This "phase" is the gas phase. At this time, even if the social space increases rapidly, the social pressure does not change significantly. But for both physical and social systems, the gas phase is less stable. Various thermodynamic parameters can easily be changed.

In the gaseous social thermodynamic system, when the social pressure changes a little, it will cause rapid changes in the social space. This situation corresponds to some relatively expansive societies. In these societies, a little change in social pressure will cause the society to slow down the increase of social pressure by expanding or shrinking social space. The stability of the entire social system cannot be maintained until the pressure between this society and other neighboring external societies is rebalanced.

However, we can also notice another situation: if a society cannot relieve the increase of internal pressure by expanding its social space, then it has to adjust its social structure to relieve this pressure. This adjustment of the social structure can be accomplished in many ways. One is through the introduction of new information technology, which can expand the information space, thereby alleviating the increased pressure caused by the entire society. Another way is to change the interaction between social individuals. For example, by enhancing the interaction between social individuals, this can also increase the critical pressure and critical temperature of the entire society, so that the entire society can exist in two different phases under normal social pressure and temperature.

6.3.6 Social thermodynamic system well below critical temperature

As can be seen from the previous examples, in many social thermodynamic systems, the critical temperature is very high. This very high critical temperature means that the system will always be well below the critical temperature.

On the other hand, from the perspective of social systems, the interaction between social individuals in many social systems is very strong, and the flow of people also has a certain integrity, just like the flow of liquids, such as European Jews, Roma and so on. This can explain that the social thermodynamic system is more similar to the liquid system.

From the solution formula (6-38) of the critical temperature, when the critical temperature is very high, it means that the interaction between individuals is very strong, or the social space of social individuals is very small.

The enhancement of interaction between individuals can be achieved in a number of ways. As in the Hun (Xiongnu) society, having a strong leader can enhance the interaction between individuals. The use of information technology to broaden the channels of information dissemination can also enhance the social interaction coupling coefficient.

On the one hand, the size of social space is limited by physical space. On the other hand, it has to do with the use of information technology. Due to the use of information technology, an individual's social space can be rapidly increased.

In general, the social temperature of a social thermodynamic system should be lower than the ambient temperature. In this way, the social thermodynamic system can efficiently obtain energy from the surrounding environment. The ambient temperature is generally around 300K. Therefore there is

$$T \ll T_C \tag{6-39}$$

We can use the inflection point equation (6-34) of the Van der Waals isotherm

$$p = \frac{a_s N^2}{V^2} - \frac{2a_s N^2 b}{V^3} \tag{6-34}$$

If calculated according to Case Study 6.2.3, the social pressure of a town before industrialization is:

$$p = \frac{N\beta}{V-b}T - \frac{a_s N^2}{V^2} \tag{6-40}$$

Substituting into the Van der Waals isotherm inflection point equation (6-33), the following equation can be obtained:

$$\left(\frac{N\beta}{V-b}T - \frac{a_s N^2}{V^2}\right)V^3 - a_s N^2 V + 2a_s N^2 b = 0 \tag{6-41}$$

After simplification we get:

$$N\beta V^3 T - 2a_s N^2 (V-b)^2 = 0 \tag{6-42}$$

or

$$\frac{V^3}{(V-b)^2}=\frac{2a_sN}{\beta T} \tag{6-43}$$

Under the condition of relatively low temperature, considering that the social space of point E satisfies $V >> b$

Taking $N = 20000$ and the social temperature of the traditional society is 272K, the above formula can be simplified as:

$$V \approx \frac{2a_sN}{\beta T} \approx 8\times10^6 m^3 \tag{6-44}$$

This corresponds to approximately 2.67 square kilometers.

Obviously, in Figure 6-3, point C is to the right of point E, so the corresponding social space is larger than $8\times10^6 m^3$, which corresponds to an area of more than 2.67 square kilometers. This shows that in such a traditional society, even if 20,000 people are distributed in an area far exceeding 2.67 square kilometers, they can form a state of coexistence of two phases. That is, the whole society is in a state of coexistence of liquid and gaseous states. Like the state of water near its boiling point. In this state, the liquid state is still maintained, but some social individuals are also "evaporated". Of course, if these 20,000 people are distributed in a larger social space, a gas social state will be formed. This state of gas society will become less stable.

Considering that this town has an area of 70 square kilometers, the actual living area of residents will still be concentrated in a relatively small area. The rest of the area is mainly farmland and mountains. Therefore, it is more realistic that the population of the town is concentrated within a few kilometers. Therefore, the entire town can basically remain in a state of coexistence of two phases of gas and liquid.

If V is not negligible with respect to b, the above equation becomes:

$$\frac{V^3}{(V-2000)^2}\approx 8\times10^6 m^3$$

obtained by numerical calculation

$$V \approx 2032m^3$$

Explain that another inflection point D corresponds to an area of about 677 square meters. This area is a bit small. Therefore, it is impossible for the whole society to completely enter the liquid or solid state without evaporation. This also reflects the fact that there are generally only two forms of condensed state and gaseous state for human society.

For the condensed state, the physical world includes both solid and liquid states. Usually the solid and liquid states of matter are very tightly bound. When a social thermodynamic system experiences a condensed state, it is usually only at the phase equilibrium line in Figure 6-1 between the condensed state and the gaseous state. The condensed state in social thermodynamic can also be divided into two states: solid state and liquid state. However, the division of these two situations is mainly based on the social mobility of a certain condensed state. If a society is basically in a very stable state of immobility, then it can be called a "solid state". And if it is in a state where it can flow at any time, it can be called "liquid state".

6.3.7 Case study: Critical temperature in Soviet society

Before and after the disintegration of the Soviet Union were two very different social states. The Soviet Union before its disintegration was a complete country, but it was a very large country. In addition, there are many republics within the Soviet Union, and the cultural differences between the republics are still relatively large. After the disintegration of the Soviet Union, a number of independent countries were formed. Although there are still some connections between these independent countries, they are no different from other independent countries in the world.

From the point of view of the social thermodynamic system, the interaction between members of the Soviet Union before the disintegration was not so strong. Although the whole country seems to be relatively strong on the surface, due to the large differences in the mutual exchanges and communication between the various republics and the ethnic groups within the Soviet Union, the relevant leaders during the Soviet Union adopted a strategy to promote the integration of the Soviet Union with a series of coercive measures. Given the size of the Soviet Union, such coercion had to be simple and effective. The interaction strength between social members formed by this simple and effective method is definitely relatively weak. Therefore, an important fact that we can know is that within the Soviet Union, the social interaction coupling coefficient between individual members of society is relatively small.

Combining with some important characteristics of the social thermodynamic system we analyzed earlier, especially some important characteristics of the Van der Waals equation, we can see that if the interaction between social members is relatively weak and the entire social space is relatively large, then there are still ways to ease social pressure. From the equation of state or Van der Waals equation of ideal gas, when the social temperature is constant, when the social space increases, the social pressure can be reduced.

On the other hand, we can also achieve changes in various thermodynamic parameters by adjusting the social interaction coupling coefficient. The reduction of the social interaction coupling coefficient mainly leads to the reduction of the critical pressure and the critical temperature.

Here we give some important social thermodynamic parameters of the Soviet Union before the disintegration:

Total population: $N = 2.87 \times 10^8$

Area: $s = 2.24 \times 10^7 \, km^2 = 2.24 \times 10^{13} \, m^2$

Calculated according to the height of 3 meters, the physical space size of the Soviet Union can be calculated as

$$V_p = 6.72 \times 10^{13} \left(m^3 \right)$$

In terms of social pressure, it mainly faces pressure from Western countries such as Europe. In terms of social space, due to the use of communication technologies such as telephones and power equipment such as internal combustion engines and motors, everyone's social space will increase. According to the estimation in Section 4.3, the collision frequency can be set at$p = 50$, so according to formula (4-21), the radius of social space occupied by social individuals in the Soviet Union is calculated to be approximately:

$$r = \sqrt{\frac{4p}{\pi n f(T)}} = \sqrt{\frac{200}{38.2\pi}} \approx 1.3(m)$$

Therefore, the Van der Waals parameter can be calculated according to formula (4-15):

$$b_S = Nv_S = Nh\pi r^2 = 2.87 \times 10^8 \times 1.7\pi \times 1.3^2 = 2.59 \times 10^9 \left(m^3\right)$$

This way we can calculate the critical volume of the Soviet Union as:

$$V_C = 3b_S = 7.77 \times 10^9 \left(m^3\right)$$

If the interaction between basic society members in the Soviet Union is only one tenth of the normal value, that is, $a_s = 1.5 \times 10^8 \, Pa \cdot m^6$, the critical pressure and critical temperature can be calculated as

$$p_C = \frac{a_s N^2}{27b_S^2} = \frac{1.5 \times 10^8 \times 2.87^2 \times 10^{16}}{27 \times 2.59^2 \times 10^{18}} = 6.83 \times 10^4 \, (Pa)$$

It can be seen that the critical pressure in Soviet society is relatively low.

$$T_C = \frac{8a_s}{27\beta v_p} = \frac{8 \times 1.5 \times 10^8}{27 \times 2.76 \times 10^4 \times 9.02} = 179(K)$$

It can be seen that there are two main reasons for the low critical temperature of Soviet society. One is the increase of social individual space. This is due to the use of information technology. On the other hand, the interaction between social individuals is weakened. This critical temperature is lower than the ambient temperature, that is to say, if the social temperature of the Soviet society is slightly increased, then the whole society can easily enter a "gas" state with only one phase. This state is not stable. As long as there is a small change in temperature, pressure, etc., it may cause a relatively large range of changes in social space.

Judging from the calculated critical temperature, as long as the social temperature of the Soviet Union is lower than 179*K*, then the Soviet society should still be in a condensed state, in which the stability of the Soviet society can be guaranteed. Below we can calculate the social temperature of the Soviet Union.

According to Example 5-8, we first calculate the basic internal energy of the individual human physiological system of the entire Soviet society. According to the calculation that the human body is entirely composed of water molecules and the average physical temperature of the human body is 310*K*, then:

$$U_p = \frac{3}{2}N\beta T = 1.5 \times 2.87 \times 10^8 \times 2.76 \times 10^4 \times 310 \approx 3.68 \times 10^{15}\,(J)$$

Then according to the per capita consumption of 1,500 kcal per day, the nutritional energy required by all individuals in the Soviet society is:

$$dE_{nutrition} = 1500 \times 4184 \times 2.87 \times 10^8 \approx 1.8 \times 10^{15}\,(J)$$

The energy obtained in the society not only meets the needs of the influence of social individuals, but also meets the energy consumed by the movement of each social individual. Assuming that the energy required for exercise is 50% of the nutrient energy absorbed every day, the temperature rise caused by the change in internal energy is

$$dT = \frac{dU}{C_v} = \frac{1.5 \times 1.8 \times 10^{15}}{7.92 \times 10^{12}} \approx 341(K)$$

It can be seen that the social temperature is higher then the environment temperature, and the change of the social temperature has exceeded its critical temperature. The kinetic energy generated by means of transport such as cars has not been taken into account here. If the impact of these kinetic energies is taken into account, the change in social temperature may be even higher. This shows that in the Soviet society it is impossible to exist in a condensed state, it can only be an unstable gaseous state.

We can also compare the social thermodynamic parameters of the United States that year. The population of the United States in 1979 was approximately 225 million. In the application of information technology, the United States has a higher degree of telephony than the Soviet Union. However, at that time, the technology of the Internet was not widely used. Therefore, the expansion of individual social space by information technology is not much larger than that of the Soviet Union. We calculate at 1.5 times the Soviet Union. That is, for a single social individual, the social space of the Americans is about 1.5 times that of the Soviet Union. Which is

$$v_U = 1.5 v_S$$

But on the other hand, the Soviet Union is information control, that is, inhibiting the exchange of information between social individuals. This resulted in less interaction between social individuals in the Soviet Union than in the United States. If the interaction between members of the Soviet Union is only one tenth of the normal, then we can assume that the information interaction coupling coefficient between members of the American society can reach ten times the normal. This is the application of information technology, which has the effect of enhancing social interaction. Which is:

$$a_U = 1.5 \times 10^{10}\, Pa \cdot m^6$$

This way we can calculate the Van der Waals constant for American society

$$b_U = N v_U = 2.25 \times 10^8 \times 1.5 \times 1.7\pi \times 1.3^2 \approx 3.04 \times 10^9 m^3$$

Then calculate the critical volume of American society as:

$$V_C = 3b_U = 9.13 \times 10^9 m^3$$

The corresponding critical pressure and critical temperature are

$$p_C = \frac{a_u N^2}{27b_U^2} = \frac{1.5 \times 10^{10} \times 2.25^2 \times 10^{16}}{27 \times 3.04^2 \times 10^{18}} = 3.03 \times 10^6 (Pa)$$

It can be seen that the critical pressure of American society is relatively high.

$$T_C = \frac{8a_u}{27\beta v_U} = \frac{8 \times 1.5 \times 10^{10}}{27 \times 2.76 \times 10^4 \times 1.5 \times 9.02} = 1.19 \times 10^4 (K)$$

It can be seen that in 1979, the critical social temperature of American society has reached 11,900*K*, which is a very high temperature. From the analysis in Section 6.3.6, it can be proved that as long as the population is properly concentrated in large cities, American society can easily enter a very stable condensed state. This is in stark contrast to the Soviet Union.

The reason why the critical temperature of American society can reach such a high temperature is mainly because the information interaction between social individuals is very strong. At that time, there was basically no speech control in the United States, which also became a reality for the development of information technology to promote communication between people. In the Soviet Union it was the opposite. This is also the reason for the low critical temperature of Soviet society. Therefore, it can be seen from this that the weakening of the information interaction between individuals is an important reason for the decrease of the critical temperature.

There are many factors that cause social pressure changes in the Soviet Union, which can be divided into two categories: internal factors and external factors. Internal factors involve the pressure of economic development and conflicts between ethnic groups. External factors include the arms race and Western economic development. At the end of the Soviet Union, after the 1970s, it belongs to the "Cold War" period. During this period, the Soviet Union faced greater internal and external social pressures. This increase in social pressure caused a further rise in the social temperature of the Soviet Union. In order to alleviate this rise in social temperature, the Soviet Union had to expand social space to achieve this. For example, in the late 1970s, the Soviet Union expanded its social space by invading Afghanistan (Gibbs, 2006). This is also a way to temporarily lower the social temperature, but in the long run, it brings more social pressure.

Of course, it can be seen from the case analysis in Section 6.2.3 that by reducing economic growth and reducing the energy input of the whole society, the temperature of the whole society can also be reduced. This may be one reason for the phenomenon of "shortage economy" in Soviet society (Nove, 1969).

6.4 The impact of information technology development on social thermodynamic parameters

6.4.1 The role of information technology

Since information technology has a huge impact on social space, the emergence of information technology will definitely have a huge impact on the state of the social thermodynamic system.

The impact of information technology on social thermodynamic parameters is mainly manifested in the following two aspects:

1. The application of information technology expands the information space. Using different information technologies to expand the information space effect is not the same. From the traditional postal system to telegraph telephones to different versions of the internet and more. At present, Web2.0 has the greatest impact on the expansion of information space.
2. Information technology affects the interaction between individual members of society. On the one hand, information technology can enhance the interaction among basic society members, which is a normal function of information technology. The reason

behind this is that "the media is an extension of the human organ" (Mcluhan, 1994). On the other hand, the application of information technology may also weaken the interaction among basic society members. In densely populated urban areas, information technology can effectively broaden the mutual contact between basic society members. But in families or some villages with a small population, the application of information technology may weaken the interaction between social individuals because social individuals have been able to communicate effectively through face-to-face means. For example, some teenagers in the family who are addicted to online games sacrifice the interaction with their relatives and friends because they frequently interact with the avatars in the game. In addition, the application of information technology may also be subject to some purposeful information regulation, thereby restricting the interaction between social individuals. For example, in the Soviet period, through media such as television and newspapers, the government could effectively restrict the communication between basic members of the society.

6.4.2 Information technology and changes in social thermodynamic system status

From the previous calculations, we can see that the introduction of information technology also leads to an increase in the social space of social individuals, and this increase in social space actually means an increase in the critical social space. At the same time, a drop in critical pressure and a drop in critical temperature are caused.

When the critical temperature of a social thermodynamic system drops, it means that the system is more likely to exceed the critical temperature and enter a state with only one phase. This phase that exceeds the critical temperature can generally be referred to as the gas phase. In daily life, we still have more intuitive understanding of the state of gas. Gases flow and diffuse more easily than condensed states like liquids and solids. Moreover, the volume change of the gas has a relatively large correlation with the pressure and temperature. As long as the pressure or temperature changes slightly, the volume of the gas will also change in a relatively large range.

But on the other hand, information technology has a great influence on the information interaction between social individuals. Different from the impact on the information space, the impact of information technology on the interaction between social individuals mainly exists in two aspects. One is to enhance the information interaction between social individuals; the other is to weaken the information interaction between social individuals. From the calculation of the critical parameters of the Van der Waals equation, it can be seen that if the information interaction increases or decreases, it has no effect on the critical social space, but it can have a very serious impact on the critical pressure and critical temperature. And this effect is in the same proportion as the change in social space.

If the introduction of information technology leads to the weakening of information interaction, it means that both the critical pressure and the critical temperature will be greatly reduced. On the contrary, if the introduction of information technology leads to the enhancement of information interaction among social individuals, then this will increase the critical pressure and critical temperature.

This also means that the change of information interaction caused by the introduction of information technology may lead to an increase in critical pressure and critical temperature, or a decrease in critical pressure and critical temperature. If the critical temperature falls, this actually means that the system is more likely to enter an unstable gaseous state, while an increase in the critical temperature means that the system is more likely to enter a very stable condensed state.

6.4.3 Case study: The impact of information technology on traditional society

In some sparsely populated rural areas, older traditional cultures may still be preserved. These traditional cultures rely on oral transmission or written records for inheritance. The application of information technology will change the way culture is passed on.

We can now do some specific calculations to analyze.

Assuming a traditional rural area with a population of N, and using traditional methods for information transmission and communication, the social space corresponding to social individuals isv_1, or the Van der Waals constant is expressed as: $b_1 = Nv_1$

Since there is already a traditional cultural enhancement of social interaction, the social interaction coupling coefficient in this region isa_1

In this way, we can establish a social equation of state for this traditional society:

$$\left(p + \frac{a_1 N^2}{V^2}\right)(V - b_1) = N\beta T \tag{5-9}$$

where p, V, T are thermodynamic parameters. And β is a constant. Through this equation we have been able to describe all the parameters of this social thermodynamic system. Its corresponding curve is also the Van der Waals curve shown in Figure 1-3. Below the critical temperature T_c, the curve possesses two "phases".

Then we feed in the data from the pre-industrial town in case study of Section 6.2.3. Although the town has thousands of years of historical and cultural heritage, and the interaction between social individuals has been strengthened, we still use the data that has been determined earlier here:

$$a_1 = a_s = 1.5 \times 10^9 \, Pa \cdot m^6$$

Considering that in the absence of new information technology applications, the information exchange among villagers is basically word of mouth, and the efficiency of written information transmission is also relatively low. Therefore:

$$b_1 = Nv_1 = 20000 \times 0.1 = 2000 m^3$$

This way we can calculate the critical parameters for the small town:

$$V_C = 3b_1 = 6000 m^3$$

$$p_C = \frac{a_1 N^2}{27b^2} = \frac{1.5\times10^9 \times 20000^2}{27\times 2000^2} = 5.56\times10^9 (Pa)$$

$$T_C = \frac{8a_1}{27\beta v_1} = \frac{8\times1.5\times10^9}{27\times2.76\times10^4\times0.1} = 1.61\times10^5 (K)$$

The larger the critical space, actually means that the social space that reaches the isotherm at a lower temperature will be larger. That is to say, in a large enough social space, the entire social thermodynamic system can enter a mixed state.

It can be seen that this is a very high critical temperature. It means that the actual social temperature will not reach this critical temperature. Considering that this small town has a history of several thousand years, so, after a long enough interaction with the environment, this small town has been able to meet the ability to maximize the use of traditional methods to obtain sufficient resources from nature and maintain social stability, so its temperature can reach the social temperature which is basically below the environment temperature. In the case study of Section 6.2.3, we have calculated that the social temperature of the small town is about $T = 272K$

so

$$p = \frac{2.76\times10^4\times20000}{V-2000}T - \frac{1.5\times10^9\times20000^2}{V^2} \tag{6-45}$$

If information technology is used, the social space of social individuals will expand rapidly, which will naturally lead to changes in critical volume, critical pressure and critical temperature, and the critical volume will increase accordingly. For the convenience of comparison, here we also assume that the introduction of information technology has no effect on information interaction. After all, thousands of years of cultural heritage can form a kind of inertia in the relationship between people. That is to say, in this traditional society, even with the most advanced information technology, the interaction between individuals can still be maintained at a relatively low level of interaction when there was no information technology application.

Which is

$$a_2 = a_1$$

After the introduction of information technology, such as web2.0, the calculation is based on a height of 1.7 meters and a radius of 10 meters. The social space of a single social individual will become

$$v_2 = 1.7 \times 3.14 \times 10^2 = 534m^3$$

therefore:

$$b_2 = Nv_2 = 20000 \times 534 = 10680000m^3$$

This way we can calculate the critical parameters for the small town:

$$V_C = 3b_2 \approx 3.2 \times 10^7 m^3$$

$$p_C = \frac{a_2 N^2}{27b_2^2} = \frac{1.5 \times 10^9 \times 20000^2}{27 \times 10680000^2} = 195(Pa)$$

$$T_C = \frac{8a_2}{27\beta v_2} = \frac{8 \times 1.5 \times 10^9}{27 \times 2.76 \times 10^4 \times 534} = 30(K)$$

It can be seen that the critical temperature becomes very low. Only those temperatures below the critical temperature will the two distinct phases appear. But in general, the entire social temperature will be significantly higher than the critical temperature. This also means that this social system always has only one gaseous phase, which can be approximately described by the equation of state of an ideal gas.

Compared with the condensed state of traditional society, the gaseous society is less stable, that is to say, with the change of external pressure, the entire social space will also undergo great changes. However, if the social temperature can be maintained at a low level of 30*K*, the critical volume of this society is still relatively large, exceeding

10 square kilometers. Therefore, to form a condensed state, the social temperature needs to be greatly reduced.

It can be seen that the role of information technology in remote rural areas has greatly reduced the critical social temperature of the rural areas, which may make these areas more unstable.

But we can also consider it from another aspect, that is, while expanding the social space of individuals, information technology can also improve the information interaction between social individuals. If the improvement of this information interaction is proportional to the improvement of social space. For example, the social space has increased by 1,000 times, and the information interaction between individuals has also increased by 1,000 times. At this time, we can see from the calculation of relevant parameters that neither the social critical temperature nor the critical pressure has changed. It is also possible to maintain a critical temperature and critical pressure as in traditional societies.

Since such a critical parameter has been able to maintain a very stable state in the entire traditional society in the past several thousand years, the same critical parameter must be obtained after the introduction of information technology. Therefore, in this case, I believe that this modern information technology-led society can also maintain a very stable state, and it may also lead to continued population growth.

This kind of information technology's changes to society, especially to rural society, now have obvious effects in some very traditional areas. For example, in Kamiyama Town on Shikoku Island, Japan, in the process of rapid industrialization, a village that will soon disappear from the map, the population here is decreasing every year. However, after the introduction of high-speed optical fiber technology and the use of advanced Internet technology, some Internet companies in Tokyo have been able to set up satellite offices in this small village, thus attracting a group of IT technicians who are willing to come to this quiet village to work. This increased the population of Kamiyama Town to about 6,200 people. This is a very large growth rate, especially for a very remote rural area, which can make the population grow rapidly, which can also play a very important demonstration effect. It is believed that it will serve as a good example for the development of rural areas in the future (Government of Japan, 2022). This also provides a positive idea for the application of information technology to reverse the population decline caused by industrialization and informatization in rural areas.

References

Burns, W. C. (1997). Global warming-the United Nations Framework Convention on Climate Change and the future of small island states. *Dickinson Journal of Environmental Law & Policy*, 6, 147.

Cahn, C., & Guild, E. (2010). *Recent migration of Roma in Europe*. Council of Europe.

Gibbs, D. N. (2006). Reassessing Soviet motives for invading Afghanistan: A declassified history. *Critical Asian Studies*, 38(2), 239–263.

Government of Japan. (2022). The digital transformation of rural Japan. https://sponsorcontent.cnn.com/int/government-of-japan/digital-transformation/?utm_medium=cnn&utm_source=nat&utm_campaign=jco22_A1_apac

Lumey, L. H., Khalangot, M. D., & Vaiserman, A. M. (2015). Association between type 2 diabetes and prenatal exposure to the Ukraine famine of 1932–33: A retrospective cohort study. *The Lancet Diabetes & Endocrinology*, 3(10), 787–794.

Mcluhan, M. (1994). *Understanding media: The extensions of man*. MIT Press.

Nove, A. (1969). *An economic history of the USSR*. IICA.

Pathria, R. K., & Beale, P. D. (2011). *Statistical mechanics*. Elsevier Science. The Boulevard, Langford Lane, Kidlington, Oxford, 0X5 1GB, UK: 1–23

Trenberth, K. E. (1981). Seasonal variations in global sea level pressure and the total mass of the atmosphere. *Journal of Geophysical Research: Oceans*, 86(C6), 5238–5246.

Trenberth, K. E., & Guillemot, C. J. (1994). The total mass of the atmosphere. *Journal of Geophysical Research: Atmospheres*, 99(D11), 23079–23088.

Verniani, F. (1966). The total mass of the Earth's atmosphere. *Journal of Geophysical Research*, 71(2), 385–391.

Werth, N. (2008). The Great Ukrainian Famine of 1932–33. *Online Encyclopedia of Mass Violence*, 1917–1922.

7

Social-oriented society

7.1 The origin of social-oriented society

7.1.1 The effect of temperature on social types

From the analysis in the previous chapters, it can be seen that the environmental temperature, that is, the atmospheric temperature, has a very direct impact on the social type. On the one hand, if the social temperature is lower than the ambient temperature, it means that the society can easily obtain energy from the environment. On the other hand, if the social temperature is higher than the ambient temperature, it means that the society will be weaker to obtain energy from the environment. This is determined by the zeroth law of thermodynamics.

But we can also know from the law of thermodynamic cycle that if enough external energy is consumed, or there is continuous external energy input, the system can also absorb energy from the low temperature environment to the high temperature environment. This is a heat engine effect.

Therefore, in an industrialized society, although industrialization causes the temperature of the whole society to be higher than the ambient temperature, because industrialization can continuously input external energy, these continuously input external energy can absorb a large amount of energy from the relatively low temperature environment and input it to the society.

Another effect of environmental temperature on society is the interaction between social individuals. The gender interaction is mainly the interaction between two different genders. This interaction between two different genders means that only two different genders can participate. From a cultural point of view, this interaction is a relatively private interaction.

While social interaction is different, social interaction is similar to gravitational interaction. In this case, the accumulation of social individuals is the basis for the enhancement of social interaction, which is just different from gender interaction.

Stronger social interaction means the aggregation of more social individuals, and the aggregation of social individuals will damage the privacy of gender interaction, thereby weakening gender interaction.

It can also be seen from the life of nomadic peoples that those who survive by nomadic way usually use the way of living in tents or yurts to graze and survive. After all, tents or yurts are easier to disassemble, which can better adapt to the transfer of these ethnic groups from one area to another at any time.

Of course, living in tents or yurts means stronger social interaction, but it also damages the privacy of life. Without enough privacy in life, the intensity of natural gender interactions is suppressed.

Due to such an effect, we can basically draw the conclusion that the colder the ambient temperature, the weaker the gender interaction. This also leads to the faster and more effective development of social-oriented society in cold regions from the perspective of global temperature distribution. In the warmer southern regions, this is mainly the formation of a family-oriented society.

From the perspective of Chinese history, since the Spring and Autumn Period and the Warring States Period, the temperature of the whole of China has been dropping continuously. Temperatures in the north are getting colder. This promotes the development of social-oriented society. We can see the relationship between the atmospheric temperature and changes in social types from the rise of Hun since the Spring and Autumn and the Warring States Period and the prosperity of ethnic minorities in northern China.

The cold climate also causes the whole society to become a society that flows and spreads easily like gas, which is the so-called nomadic society. Due to the cold climate and the slow growth rate of plants, they cannot graze sheep in one area for a long time. At this time, the whole society has to continuously migrate with the change of grassland vegetation. This is completely different from the way of life in the warm climates of the South. The weather in the south is warmer and more suitable for the rapid growth of plants, so once a society in the south is formed, it can continue to grow various food crops in an area to survive.

The warm climate also ensures the privacy of social individuals, thereby enhancing gender interaction. More importantly, this stability of southern society also contributed to the protection of private property.

On the other hand, the social-oriented society in the cold northern regions, because of the nomadic way of living in these regions, no piece of land is privately owned by an individual or a member of society. An important foundation of social-oriented society is the absence of private property. This nomadic life is fully guaranteed. This is also an important reason why it is easy to form a social-oriented society in the north.

In this way, we can conclude that the coldness of the climate has forced humans living in the higher northern latitudes of the earth to gather together, and this social gathering is at the expense of human privacy. This sacrifice of privacy also means the weakening of gender interaction. The weakening of gender interaction is the most important foundation for the formation of social-oriented society. On the other hand, nomadic life leads to social-oriented society in which social individuals have no fixed place to live, so there is no concept of private property. Therefore, public ownership is a distinctive feature of social-oriented society.

7.1.2 History

The difference in the type of society caused by the high and low temperature is an important example of the physical condition determining the social state. But on the other hand, we can also notice that human society has cultural heritage. Although the temperature of the environment will change greatly, the cultural inheritance of human society can break through the limitation of physical temperature and maintain the type of society that it has.

For example, in Europe, although the latitude is similar to that of northern China, because European culture was inherited from ancient Greece and ancient Rome with a subtropical climate, many European society has also integrated many family-oriented society factors into the later historical development process. As a result, European society has become a typical balanced society with both family-oriented society characteristics and certain social-oriented society characteristics.

In the ancient Hun society in northern China, the Hun society close to the south was exposed to more culture of the Central Plains society in China. This makes it easier for the southern Hun society to inherit the culture of a family-oriented society and transform it into a family-oriented society. However, due to historical reasons and the influence of objective physical conditions, it is difficult for the Hun society in the north to be influenced by the family-oriented society culture in the south. Ultimately, the northern Hun has been flowing across Eurasia for hundreds of years. And caused changes in the social structure of the entire Eurasian continent.

Compared with the development of ancient nomadic peoples in Europe, the nomadic peoples in the cold regions of northern Asia have a longer history and developed faster. If we compare the temperature changes in history, we can also notice that the development process of nomadic peoples in northern Asia is consistent with the temperature changes in the entire northern latitudes. That is to say, when the temperature in the northern latitudes and high latitudes is relatively cold, the nomadic peoples in northern Asia develop faster.

Before the Spring and Autumn Period and the Warring States Period in ancient China, the temperature of the entire northern latitudes of the earth was relatively high. At this time, although there were a considerable number of nomads in high latitudes, the differences between these nomadic societies and those at lower latitudes were not very large. They are not even recorded much in the history books.

However, in the Spring and Autumn Period and the Warring States Period, we can see from the changes in global temperature in history that the global temperature at this time has a relatively large drop. Comparing with history, we can find that in China, the prosperity of Hun started from the Spring and Autumn Period and the Warring States Period.

Of course, the rise of Hun is also related to the change of another thermodynamic parameter. Before the Spring and Autumn Period and the Warring States Period, China was a unified dynasty, and the unified dynasty was very powerful, which caused the nomads in the north to face great pressure, which made the survival of these nomads also have great problems. From the point of view of the principle of social interaction, the powerful Zhou Dynasty was very attractive to the surrounding small societies, which made it easier for these small societies to be quickly integrated into the Zhou Dynasty.

However, in the Spring and Autumn Period and the Warring States Period, the entire Zhou Dynasty was split into many small countries, so that the entire Central Plains society was divided, and there was no greater pressure on the north. For a social thermodynamic system, if the external pressure decreases, it means that the thermodynamic system may expand rapidly. The rapid expansion process, if there is no external input energy, is an adiabatic expansion process.

For human society, the historical changes of decades can be regarded as a process of adiabatic expansion. For these nomads in northern China, the external pressure suddenly decreased, resulting in adiabatic expansion, which directly led to a rapid decrease in the social temperature of the nomads throughout northern China. The rapid decrease in social temperature means that the social thermodynamic system formed by these nomads will more easily obtain energy from the environment. The input of more energy has promoted the rapid development of these nomadic societies.

The rapid development of these nomadic societies directly led to the appearance of the first emperor of the Hun society, Touman Chanyu (Vaissière, 2006). The appearance of Touman Chanyu was a very important event because Touman Chanyu was able to bring together the Hun tribes that were scattered in different geographical locations. The concentration of the Hun tribe makes it possible to concentrate more social individuals in a relatively small area. From the calculation of various equations of social thermodynamic, we can see that when the number of social individuals increases, it will directly lead to a rapid increase in the heat capacity of the society. The rapid increase in heat capacity means that the social thermodynamic system can absorb more energy, but the social temperature does not rise rapidly. This gives the society greater stability.

The appearance of Modu Chanyu makes the cohesion of the Hun society even closer (Kim, Brenner, Mair, & et al, 2010; Gunchinsuren, 2017). By the end of the Spring and Autumn Period and the Warring States Period, the strength of the Hun society had seriously affected the stability of Chinese society. Therefore, in the time of Qin Shihuang, the scattered city walls of the past countries finally began to be connected together, forming what is now called the Great Wall.

Of course, the emergence of a unified China will also cause a very large social pressure to the Hun society. The appearance of this pressure has caused the thermodynamic system of the Hun society to be compressed. Of course, because the compression time is relatively short, it can also be regarded as an adiabatic compression process. Contrary to adiabatic expansion, the process of adiabatic compression leads to a decrease in social space on the one hand, and an increase in social temperature on the other hand. Although in a physical thermodynamic system, this compression is more likely to form a condensed state, that is, at a relatively high temperature, because this compression can enhance the social interaction coupling coefficient, so the critical temperature can be increased. But the difference between social thermodynamic system and physical thermodynamic system is that the level of social temperature is directly related to the ambient temperature. If this adiabatic compression process causes the temperature to rise, the social temperature will be significantly higher than the ambient temperature. This will lead to a reduction in the Hun society's ability to harvest energy from the environment. Once the Hun society cannot obtain enough energy from the environment, it faces an existential crisis. And this existential crisis is one of the most important factors leading to the weakening of interaction within Hun society.

The substantial weakening of the interaction within the Hun society has led to a split within the Hun society. By the Han Dynasty, the Hun society had split into the Southern Hun and the Northern Hun, two societies with very different social structures. Due to the influence of Chinese culture, South Hun is more inclined to family-oriented society. The North Hun still retains the original social-oriented society structure.

From a social thermodynamic perspective, South Hun has become a family-oriented society. In terms of social form, there is a very obvious difference with North Hun. At this time, the South Hun, just like the Central Plains before, created a relatively large social pressure on the North Hun. The formation of this relatively large social pressure actually resulted in the compression of the social space in the North Hun and increased social pressure, and lead to an increase in social temperature. If the social temperature rises more seriously, it will lead to the weakening of the ability of the entire North Hun society to obtain energy from the environment. This directly led to the conflict between the North and the South, and the South, relying on the strong economic strength of the Central Plains, eventually drove the North out of Northeast Asia and into Central Asia.

After entering in Central Asia, the North Hun strength is already very weak, but the vast grasslands in Central Asia have brought very good conditions to the recovery of the North Hun. In particular, the development of civilizations in Central Asia and Europe is still far behind that in eastern Asia. The social pressure faced by the Hun society has plummeted, and the Hun society can achieve rapid development similar to that in the Spring and Autumn Period and the Warring States Period in China. Finally in the 5th century AD, under the leadership of Attila, Hun re-emerged.

As in Northeast Asia, Hun is not a nation. It is mainly a form of social-oriented society, so there are actually many ethnic components in the Hun society. The nomadic people scattered all over north Europe adopted the organizational form of the Hun society in North Asia, strengthening the interaction between social individuals. The increase of the population and the expansion of the social space make the social temperature decrease significantly. The increase of interaction between social individuals and the enhancement of the coupling coefficient makes the critical temperature also have a very high rise. These all help to promote the formation of a very stable cohesion of the Hun society in Europe. The social-oriented society in this condensed state is naturally much stronger than the Roman Empire of that time.

7.1.3 Case study: Thermodynamic parameters of Hun society

Suppose there were already a million people in northern China before Touman Chanyu appeared. These populations are scattered over a million square kilometers of grassland. And because the population is so dispersed, there is little interaction with each other. In this way, various social parameters of this society can be calculated.

First of all, because the main food of the northern nomads is meat. Therefore, the calorie obtained is higher than that of the ethnic groups in the south, and of course lower than that of modern society. Here, each social individual is selected to obtain 1,400 kcal of calories per day. Therefore, it can be calculated that the total energy consumed by the social individual of the northern nomads before the emergence of Touman Chanyu is:

$$dE_{nutrition} = 1400 \times 4184 \times 1 \times 10^6 \approx 5.82 \times 10^{12} (J)$$

Considering that other energy absorbed by the human body accounts for 50% of the total nutrition (Such as herding sheep, food processing and so on), the social temperature of the Hun society is:

$$dT = \frac{dU}{C_v} = \frac{dU}{N\beta} = \frac{1.5 \times 5.82 \times 10^{12}}{10^6 \times 2.76 \times 10^4} = 316(K)$$

It can be seen that this social temperature is equivalent to 43 degrees Celsius. If it is in the warm and humid regions of the south, by reducing the nutrient intake and the proportion of other energy, the social temperature can still be lowered below the ambient temperature. But at the time of the ice age, the average temperature across the north was around 0 degrees Celsius. It is really too difficult for the Hun society to reduce the social temperature through its own methods. This eventually led to the occurrence of frequent famines in northern nomadic areas.

On the other hand, we can also calculate the critical parameters of the entire Hun society. Although the Hun people did not use information technology, they used horses to transmit information more quickly. Therefore, the social space occupied by a single individual will be slightly larger. Can be set to:

$$v_{hun} = 0.1m^3$$

Then

$$b = 1\times 10^6 \times 0.1 = 1\times 10^5 m^3$$

Thus we get the critical social space of the Hun society as:

$$V_C = 3b = 3\times 10^5 m^3$$

Considering that Hun has not formed a whole, the social interactions are very weak, and even some tribes often confront each other. Therefore, in this decentralized state, its coupling coefficient can be taken as $a_{hun} = 1.5\times 10^6 Pa\cdot m^6$. That is, it is about one thousandth of the normal social interaction. The critical pressure is:

$$p_C = \frac{a_{hun}N^2}{27b^2} = \frac{1.5\times 10^6 \times 1\times 10^{12}}{27\times 10^{10}} = 5.56\times 10^6 Pa$$

Substituting into formula (6-38), we can find the critical temperature as

$$T_C = \frac{8a_{hun}}{27\beta v_{hun}} = \frac{8\times 1.5\times 10^6}{27\times 2.76\times 10^4\times 0.1} = 161(K)$$

It can be seen that this critical temperature is very low. It was also significantly lower than the social temperature of the Hun society at that time. Therefore, in this state, it is impossible for the Hun society to form a condense state. At this time, we can basically use the equation of state of the ideal gas for analysis the northern nomads. According to the ideal gas equation of state, we can calculate the social pressure of the entire Hun society as:

$$p = \frac{N\beta T}{V} = \frac{10^6\times 2.76\times 10^4\times 316}{10^{6+6}\times 3} = 2.90(Pa)$$

It can be seen that the external pressure on the entire Hun society was also very small. Because the farming peoples of the south are also not very interested in these cold regions of the north.

If the social temperature is too high, it can still be fine-tuned by changing external pressures and social spaces, etc. For example, in the case of constant pressure, according to Charlie's law, the method of reducing social space can appropriately lower the social temperature. And if the social space is kept unchanged, according to Gay-Lussac's law, by reducing the external social pressure, the social temperature can also be lowered.

Now let's calculate the changes of some thermodynamic parameters after the appearance of Touman Chanyu.

Because of the emergence of a leader, the nomadic tribes on the entire steppe were unified to form a complete society. In this society, the interaction between social individuals can reach a normal level. That is, $a_{hun} = 1.5\times 10^9\, Pa\cdot m^6$

In this way we can calculate that the new critical parameters of the Hun society under the leadership of Touman Chanyu will become:

$$V_C = 3\times10^5 m^3$$

$$p_C = \frac{a_{hun}N^2}{27b^2} = \frac{1.5\times10^9\times1\times10^{12}}{27\times10^{10}} = 5.56\times10^9 Pa$$

$$T_C = \frac{8a_{hun}}{27\beta v_{hun}} = \frac{8\times1.5\times10^9}{27\times2.76\times10^4\times0.1} = 161000(K)$$

It can be seen that the critical temperature of the entire Hun society became very high. This also means that the Hun society will easily enter a condensed state, such as hundreds of thousands of people gathered in a relatively small social space.

After Touman Chanyu's leadership, the population of the entire Hun society will be concentrated. The benefit of this population concentration is that on the one hand, it strengthens the interaction between social individuals, and on the other hand, it can form a fluid society like a liquid. In the process of flow, the integrity of the social structure can always be maintained. Calculated according to the social temperature of 316K, and substituting into formula (5-11), we can get:

$$s = \frac{V}{3} = \frac{a_{hun}N}{3\beta T} = \frac{1.5\times10^9\times10^6}{3\times2.76\times10^4\times316} \approx 5.73\times10^7\left(m^2\right)$$

This is equivalent to nearly 60 square kilometers. That is to say, although in the Spring and Autumn Period and the Warring States Period, the grassland area occupied by the Huns reached one million square kilometers. But in fact the main population is still concentrated in about 60 square kilometers. This is equivalent to the area of several Hun cities that year. This is also supported by archaeological evidence. At present, the archaeological findings on the ruins of the capital city "Tongwancheng" (Obrusanszky, 2009; Cui & Chang, 2013) abandoned by the Hun in those days show that the area of the Hun's capital was about ten square kilometers, and its population reached about 100,000 at its peak.

In this condensed state, we can also calculate its internal pressure:

$$p_i = \frac{a_{hun}N^2}{V^2} = \frac{1.5 \times 10^9 \times 10^{12}}{1.72^2 \times 10^{16}} \approx 5.1 \times 10^4 \,(Pa)$$

It can be seen that the internal pressure increases significantly.

If the Hun society is unwilling to appropriately reduce nutrient intake, it is also unwilling to reduce the proportion of other energy absorption, such as giving up horse riding. The Hun society could also gain additional energy indirectly through the farming society in the south. This also led to the frequent invasion and harassment of the southern society by the northern Hun, so that in order to prevent the Hun invasion, the Great Wall of China had to be built in southern China. After the rise of Mongol society, the Chinese Ming Dynasty fortified the Great Wall.

Of course, there is another factor that can also reduce the temperature of society. This is the fact that the Huns gathered together, and the property was publicly owned, which could also effectively distribute energy. This can improve the efficiency of various energy utilization. Thereby reducing the internal energy of the whole society, and further reducing the social temperature.

According to Example 5-8, we first calculate the basic internal energy of the human body physiological system of the Hun society in this era. According to the calculation that the human body is entirely composed of water molecules and the average physical temperature of the human body is 310*K*, then:

$$U_p = \frac{3}{2}N\beta T = 1.5 \times 1 \times 10^6 \times 2.76 \times 10^4 \times 310 \approx 1.28 \times 10^{13} \,(J)$$

Next, we can calculate the state parameter changes of the Huns before and after Touman Changyu took power. In the first, the internal energy of the Hun society is:

$$U = U_p + 1.5dE_{nutrition} = 1.28 \times 10^{13} + 1.5 \times 5.82 \times 10^{12} \approx 2.15 \times 10^{13} \,(J)$$

We can also calculate the entropy and free energy of the Hun society in the Autumn and Warring States Period.

$$S = N\beta\left\{\frac{3}{2}lnT + ln\frac{V}{Nn_p} + \frac{3}{2}\left[\frac{5}{3} + ln\left(\frac{2\pi mk_B}{h^2}\right)\right]\right\}$$
$$= 10^6 \times 2.76 \times 10^4 \times \left\{\frac{3}{2}ln316 + ln\frac{3\times10^{12}}{10^6\times2\times10^{27}} + \frac{3}{2}\times\left[\frac{5}{3} + ln\left(\frac{2\pi mk_B}{h^2}\right)\right]\right\}$$
$$\approx 7.76\times10^{11}\left(J\cdot K^{-1}\right)$$

In this way, the Helmholtz free energy of the Hun society can be obtained as:

$$F = U - TS = 2.15\times10^{13} - 316\times7.76\times10^{11} \approx -2.24\times10^{14}\,(J)$$

Before Touman Changyu took power

$$pV = 2.9\times3\times10^{12} \approx 9\times10^{12}\,(J)$$

In this way, the Gibbs free energy can be obtained as

$$G = F + pV = -2.15\times10^{14}\,(J)$$

After Touman Changyu took power, the Huns were divided into two parts, and one part continued to live a nomadic life as before. But most of them started to gather around Touman Changyu. So it can be roughly estimated that about 400,000 are still living as before. Under the leadership of Touman Changyu, 600,000 people gathered in a social space of about $57.3km^2$. In this way, we can calculate the state parameters after Touman Changyu took power.

The first is entropy:

$$S = S_1\left(V_1, N_1\right) + S_2\left(V_2, N_2\right) = 3.2\times10^{11} + 3.12\times10^{11} = 6.32\times10^{11}\left(J\cdot K^{-1}\right)$$

It can be seen that after Touman Changyu came to power, the entropy of the entire social system decreased. This shows that society has become more orderly. Of course, this ordering requires additional energy consumption.

In this way, the Helmholtz free energy of the Hun society after Touman Changyu took power can be obtained as:

$$F = U - TS = 2.15\times10^{13} - 316\times6.32\times10^{11} \approx -1.78\times10^{14}\,(J)$$

Since after Touman Changyu took power, the Huns split into two populations. There will be less external pressure on the nomadic populations who still live their old lives. Then

$$p_1V_1 = 0.4\times2.9\times3\times10^{12} \approx 3.6\times10^{12}\,(J)$$

And the pressure of the population gathered around Touman Changyu increased and the social space decreased. Then

$$p_2V_2 = 5.1\times10^{4}\times1.72\times10^{8} = 8.77\times10^{12}\,(J)$$

In this way, the Gibbs free energy can be obtained as

$$G = F + p_1V_1 + p_2V_2 = -1.66\times10^{14}\,(J)$$

7.1.4 Modern social-oriented society

In modern society, modern social-oriented society also appears in the northern part of the earth, the most typical of which is the social-oriented society of the Soviet Union.

Since the Soviet society appeared in the 20th century, all kinds of materials are very complete, so this provides us with a lot of important materials for the study of social-oriented society. This provides an important basis for us to study the morphological structure of social-oriented society and the interaction between social individuals in a more systematic and in-depth manner.

From the existing data, we can see that the social formation of the Soviet Union has the following very important characteristics.

One is the public ownership of property, which means that social individuals own as little private property as possible, including state land, as well as some large facilities, enterprises and products, etc., which are all owned by the state. The necessary needs of social individuals are mainly distributed according to the state.

The second is the inhibition of gender interaction. In the Soviet society, there were very strict restrictions on various works of art, especially in terms of ideology, which led to the fact that in terms of artistic creation, many artists could only create along a certain direction. When it comes to sexual content, there will be very strict restrictions. This also accords with some important basic characteristics of social-oriented society. Because in social-oriented society, the main emphasis is on social interaction. Thereby weakening the gender interaction within the family, making the family relationship not as strong as the family-oriented society.

The third is to develop the military. The requirements for economic development are all for military service. This is basically consistent with Hun's social and historical development. The economic strength of the ancient Hun society was relatively weak, but because the Hun society could apply limited economic resources to the military, the military strength of the Hun society exceeded that of the southern family-oriented society in many periods. During the Soviet period, its defense expenditure even reached about 15% of the GDP of the entire national economy (Firth & Noren, 1998). However, some experts in related fields pointed out that in fact, the uncertainty of the Soviet Union's GDP in that year was very large (Engerman, 1999). Judging from the economic development after the disintegration of the Soviet Union, the actual economic scale of the Soviet Union is much smaller than its official announcement.

The fourth is the society in the social-oriented society. The social interaction between individuals is mainly maintained by the ability of leaders. In Europe, after the death of Hun leader Attila, the entire

Hun kingdom began to fall apart and eventually decline (Thompson & Drinkwater, 2015; Kelly, 2010). The same is true in the Soviet Union. After the death of Brezhnev, the leader of the late Soviet Union, the entire Soviet Union seemed to be leaderless, and eventually the Soviet Union disintegrated shortly after his death. But in family-oriented society, the role of this leader seems to be less important. In ancient China, as well as in the Roman Empire, the emperor used a hereditary system. Although the next generation does not have much ability, especially the ability to govern the country may also be relatively weak. But the whole country is still able to maintain a very stable situation. Some studies have now shown that the change of dynasties in ancient China is mainly related to changes in temperature. The demise of the Roman Empire was mainly related to the development of social-oriented society in the north, such as the rise of the Hun Empire and so on. In addition, we can also notice the phenomenon that although there are many hereditary princes and nobles in Europe, the existence of these princes and nobles basically has little effect on the country.

Of course, although there are many similarities between the Soviet society and the ancient Hun social-oriented society in many respects, we can also notice that the Soviet social-oriented society was in the period of industrialization, and information technology was also developing very rapidly. Industrialization means that energy has been efficiently harvested in Soviet times. People's living environment is no longer gathered in a short tent. In this way, many people's concept of private property has been strengthened. Gender interaction was also enhanced. This is also the reason why the Soviet society had to adopt a series of active measures to suppress gender interaction, such as in the media communication of the public, restricting the description of sexuality in literary works, and art must obey the needs of politics, etc. (Fox, 1976). In addition, it can be seen from some documents in the 1950s that since the establishment of the Soviet Union, the traditional family structure has been greatly reformed. Sexual relationships like those in the United States and other Western countries are labeled as "dirty" and "depraved". This actually denies the complexity and diversity of gender interactions, and attempts to "purify" sexual interaction in a very simple way (Sandomirsky, 1951). Of course, the purpose of this is to maintain the integrity of the social-oriented society.

The development of information technology has also increased the social space of individuals. But technology has two sides after all, it is a double-edged sword, it can use information technology to enhance social interaction, on the other hand, it can also use information technology to weaken social interaction. In the Soviet period, this aspect was very effective, including the use of radio and television and other media to restrict the thinking of social individuals, thereby restricting the adequate exchange of information between social individuals. And further weaken the social coupling coefficient between social individuals.

The weakening of the interaction between social individuals and the increase of the social space possessed by social individuals all lead to a decrease in the critical temperature of the social thermodynamic system. At the same time, the social temperature is rising rapidly. Naturally, Soviet society tends to become an unstable gaseous state, and it is difficult to enter a condensed state. This unstable gas state means that Soviet society is easy to expand and contract. When the Soviet Union's military strength is relatively strong and the internal social temperature is not very high, the Soviet society has a relatively strong ability to expand. However, when the social and economic development of the Soviet Union is not enough, Soviet society can easily shrink.

7.1.5 Case study: Thermodynamic parameters of Ukraine before and after the annexation of Crimea

Russia annexed Crimea in 2014. This had a great impact on Ukrainian society. Here, from the perspective of social thermodynamics, the thermodynamic parameters before and after the annexation of Crimea are calculated and compared. Perhaps it can be seen that after some social thermodynamic parameters of a social-oriented society change, other thermodynamic parameters are affected.

First of all, when including Crimea, Ukraine has an area of 603,628 square kilometers and a population of 43.04 million. Crimea covers an area of 26,000 square kilometers and has a population of 1.91 million.

It is calculated according to the nutrition that each social individual obtains 1500 kcal calories.

$$dE_{nutrition} = 1500 \times 4184 \times 4.3 \times 10^7 \approx 2.7 \times 10^{14} (J)$$

If the proportion of other energy consumed by modern society is calculated at 60%, the social temperature in Ukraine can be calculated as:

$$dT = \frac{dU}{N\beta} = \frac{1.6 \times 2.7 \times 10^{14}}{4.3 \times 10^7 \times 2.76 \times 10^4} = 364(K)$$

This temperature is significantly higher than ambient temperature, which is a result of industrialization leading to an increase in nutrition and the additional energy expended for each individual.

Of course, as the second largest country in Europe, it has a vast territory. Therefore, the population is mainly concentrated in the cities. Kyiv is about 800 square kilometers, and the second largest city, Kharkiv, has an urban area of about 350 square kilometers. The two big cities are home to about 4.4 million people. Therefore, it can be estimated that the population gathering density of Ukraine is about 0.38 million people per square kilometer. In this way, it can be estimated that the area where the Ukrainian population is concentrated is approximately

$$s = \frac{4.304 \times 10^7}{3800} = 11300\left(km^2\right)$$

According to the equation of state of an ideal gas, we can calculate the external pressure on Ukrainian society:

$$p = \frac{N\beta T}{V} = \frac{4.3 \times 10^7 \times 2.76 \times 10^4 \times 388}{11300 \times 10^6 \times 3} = 1.27 \times 10^4 (Pa)$$

Considering that after the collapse of the Soviet Union, Ukraine has not had a strong leader, so the coupling coefficient of social interaction will be relatively small. Calculated as an order of magnitude smaller than the normal coupling coefficient, the social interaction coupling coefficient of Ukrainian society before the annexation of Crimea is

$$a_{Ukraines} = 1.5 \times 10^{8} \, Pa \cdot m^{6}$$

So the internal pressure is

$$p_i = \frac{a_{Ukraines} N^2}{V^2} = \frac{1.5 \times 10^{8} \times 4.3^2 \times 10^{14}}{\left(11300 \times 10^{6} \times 3\right)^2} = 247\,(Pa)$$

It can be seen that the internal pressure of Ukrainian society in this state is far less than the external pressure, which is actually in a gaseous state. It can be approximated by the ideal gas equation of state.

In addition, we can also calculate the critical parameters of Ukrainian society.

Here, according to the information space $v_u = 0.5m^3$ of a single social individual before the annexation of Crimea, the critical social space of Ukraine is

$$V_c = 3 \times 4.3 \times 10^{7} \times 0.5 \approx 3 \times 2.2 \times 10^{7} = 6.6 \times 10^{7} \left(m^3\right)$$

It can be seen that the critical social space is small.

$$p_C = \frac{a_{Ukraines} N^2}{27 b^2} = \frac{1.5 \times 10^{8} \times 4.3^2 \times 10^{14}}{27 \times \left(2.2 \times 10^{7}\right)^2} \approx 2.28 \times 10^{7}\,(Pa)$$

$$T_C = \frac{8 a_{Ukraines}}{27 \beta v_u} = \frac{8 \times 1.5 \times 10^{8}}{27 \times 2.76 \times 10^{4} \times 0.5} = 3302(K)$$

It can be seen that the critical temperature and critical pressure are still relatively high.

After the annexation of Crimea, the total number of social individuals was reduced to $N = 4.113 \times 10^7$. The social space was reduced to

$$s = \frac{4.113 \times 10^7}{3800} = 10800\left(km^2\right)$$

After the annexation of Crimea, there was an urgent need for Ukrainians to obtain more information, which led Ukrainians to use as many information technologies as possible to obtain the required information. Since the information technology utilization rate of social individuals has been improved, the calculation here is based on the 10-fold increase in the information space of a single social individual. That is $v_u = 5m^3$. Another effect is that the social interaction within Ukrainian society is greatly improved. This is also calculated according to an increase of an order of magnitude. Then the interaction coupling coefficient between individuals in Ukrainian society after the annexation of Crimea becomes

$$a_U = 10a_{Ukraines} = 1.5 \times 10^9 \, Pa \cdot m^6$$

So that we can calculate the new critical parameter becomes

$$V_c = 3 \times 4.1 \times 10^7 \times 5 \approx 6.2 \times 10^8 \left(m^3\right)$$

$$p_C = \frac{a_U N^2}{27b^2} = \frac{1.5 \times 10^9 \times 4.3^2 \times 10^{14}}{27 \times \left(2.1 \times 10^8\right)^2} \approx 2.28 \times 10^6 \,(Pa)$$

$$T_C = \frac{8a_U}{27\beta v_u} = \frac{8 \times 1.5 \times 10^9}{27 \times 2.76 \times 10^4 \times 5} = 3302(K)$$

It can be seen that the critical pressure has decreased while the critical volume has increased. But the critical temperature did not change.

From this result, it can be seen that after the annexation of Crimea, Ukrainian society became somewhat sensitive to changes in external pressures. That is, the change in external pressure will now be more likely to approach or even exceed the critical pressure. The same effect was evident in other republics of the former Soviet Union.

Since the structure of energy consumption in Ukrainian society has not changed, the social temperature and external social pressure remain the same as before the annexation.

Let's take a look at the changes in internal pressure, namely:

$$p_i = \frac{a_U N^2}{V^2} = \frac{1.5 \times 10^9 \times 4.1^2 \times 10^{14}}{\left(10800 \times 10^6 \times 3\right)^2} = 2470(Pa)$$

It can be seen that the internal pressure has also increased tenfold. Compared with the pressure of 12700*Pa*, this has reached a ratio of 20%. Therefore, the influence of internal pressure on external pressure becomes more obvious.

In general, before the annexation of Crimea, the social interaction coupling coefficient was relatively weak due to the large differences in views held by social individuals and ethnic groups within Ukraine. But this weak coupling factor can at least preserve the apparent mutual cooperation and state integrity of the entire social system. After Russia annexed Crimea, the coefficient of social interaction coupling increased significantly. Once the coefficient of social coupling increases, it also means that the opposition of different viewpoints becomes more dominant. The level of conflict may thus be exacerbated. Many conflicts that were originally just at the level of verbal discussions may be directly implemented into actions.

On the other hand, after the annexation of Crimea, the internal pressure of Ukrainian society increased enough to obviously affect the external pressure. That is, after divisions within Ukrainian society cause some external pressure to feel that things may be moving in the opposite direction of their expectations, corrective measures may be taken. This may be the reason for Russia's direct intervention in Ukraine's internal affairs by invading Ukraine in 2022. This in turn led to direct Western military aid to Ukraine.

7.2 Interaction method

7.2.1 Gender interaction in social-oriented society

It has been shown in this book many times that gender interaction in social-oriented society is relatively weak. In the early stage of social-oriented society development, this relatively weak gender interaction was mainly caused by the different lifestyles from family-oriented society due to the influence of ambient temperature. Therefore, this relatively weak gender interaction exists naturally. However, with the development of society, the advancement of technology, especially the arrival of the industrialized society, people can effectively obtain the energy of the environment no matter what the ambient temperature is. This makes the weak gender interaction foundation that social-oriented society relies on no longer exists. But the culture of human society has a strong inheritance. This also leads to weak gender interaction, changing from the original natural existence to the mandatory requirements in modern social-oriented society.

One of the more typical examples is that the Russian bishop Kirill hates homosexuality very much. He even thought that "gay pride parades were part of the reason for Ukraine war". (Gallagher, 2022)

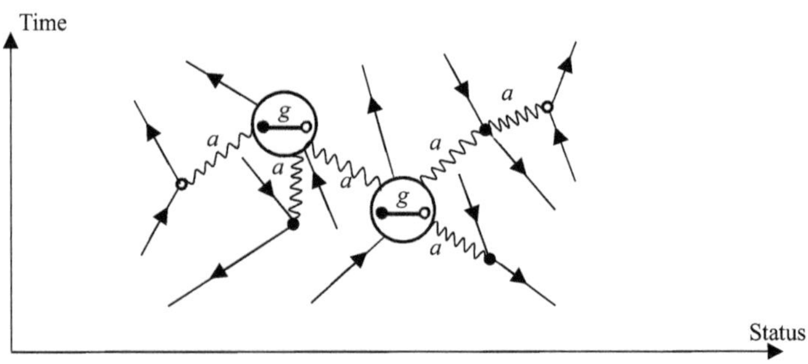

Figure 7.1. The Feynman diagram of Social Oriented Society

Therefore gender interaction is only a tool to satisfy specific desires in social-oriented society. Since gender interaction only plays an instrumental role, it means that it should serve the development needs of the entire social-oriented society, and it should be subordinate to the requirements of social interaction.

The instrumental nature of gender interaction means that children born from gender interaction are only constituent members of society. In the family, the relationship between parents and children should be the same as the relationship between other members of the society. This leads to a weaker social interaction between parents and children in a social-oriented society.

We can use Feynman diagram to represent the relationship between gender interaction and social interaction in social-oriented society.

Contrast the social-oriented society shown in Figure 7-1 with the family-oriented society shown in Figure 2-5. It can be found that in social-oriented society, gender molecular is relatively small. Therefore, its interactions with other gender molecules and other basic society members are basically equal. This also makes the information interaction between basic society members in social-oriented society stronger than in family-oriented society. Therefore, we can see that in a social-oriented society, two social individuals with different genders can form a basic society member. This also leads to a very simple social structure of social-oriented society.

In addition, from the perspective of gender interaction, it is very similar to electromagnetic interaction. Macroscopic electromagnetic interaction has a strong macroscopic effect, and can even produce an effect equivalent to that of macroscopic gravitational interaction in certain situations. Such as magnetic levitation phenomenon and so on.

Therefore, theoretically, gender interaction can also produce a more macro-social effect equivalent to social interaction. However, once the macro effect of gender interaction is revealed, it may have an impact on social interaction. Influence the intensity of social interaction. This is especially evident in family-oriented society. Homosexuality may be one of such macro gender interactions. This is also an important reason why Russian bishop Kirill strongly opposes homosexuality. I believe that in his thinking, he has a very deep understanding of the attributes of Russia's social-oriented society. Noticed that gender interaction has an important influence on the stability of the entire Russian social-oriented society. If it can limit, or even prohibit the macroscopic effects of these gender interactions. It should be able to maintain social interaction in a stronger social-oriented society. In this way, the entire Russian society is maintained in a relatively stable condense state.

7.2.2 Social interaction in social-oriented society

Since it is a social-oriented society, as the name implies, it means that in such a type of society, social interaction is the main form.

Social interaction is essentially the interaction between social individuals or basic society members. This interaction is of course produced by the exchange of information in this type or society. So social interaction is much simpler than gender interaction in social-oriented society. After all, in gender interaction, in addition to the information in various languages, there are other more information, including the transmission of sex hormones and so on.

Social interaction in social-oriented society mainly occurs between social individuals and social individuals, as well as between social individuals and basic society members such as gender molecular.

Molecular structure is relatively simple due to gender in basic society members. Unlike families in family-oriented societies, the structure of the macromolecular is complex. Therefore, when analyzing the thermodynamic problem of social-oriented society, gender molecular can be directly treated as a social individual.

Interaction in social-oriented society mainly depends on the transmission of information. The greater the total amount of information transmission, the stronger the social interaction will be. This is similar to gravitational interaction in physics. That is to say, the information interaction has a cumulative effect. This also means that once a social-oriented society reaches a certain scale to form an information blackhole, the entire social-oriented society can have a huge attraction to other individual members of society outside.

But social-oriented society has a flaw. Before the industrialization of society, the energy acquisition efficiency of the whole society was very low. At the same time, it is subject to very backward information technology, which also means that the efficiency of information transmission is also very low. Therefore, social interaction in social-oriented society is mainly influenced by leaders. If the role of the leader in a social-oriented society is very strong, he can enhance the social interaction in this social-oriented society. Conversely, social interaction decreases rapidly, which may also be the reason why the social scale of the Hun society in history cannot reach the huge scale of the family-oriented society in the Central Plains of China.

After the development of information technology, the scale of social-oriented society has expanded rapidly. This was mainly formed in the medieval Mongol-dominated Eurasian empires. In the 16th century, due to the development of transportation technology and the widespread use of gunpowder, the efficiency of human energy acquisition has been rapidly improved. In this way, even if the social temperature rises, it is easier to obtain energy from the environment. The efficient transmission of information through such means as post stations enables the social-oriented society to transmit social interaction to a long distance. This is the reason why social-oriented society can continue to expand. In contrast, the interaction range of covalent bonds or ionic bonds in family-oriented society is very short.

The analogy is that in the 20th century, the development of communication technologies such as various telephones and telegraphs made the transmission of information more distant. The application of this information technology, if not interfered, can significantly enhance social interaction. Let the distance of social interaction become more distant. Under the application of this information technology, the emergence of a huge social-oriented society like the Soviet Union seems to have become a historical necessity.

The increase in the total amount of information in the entire Soviet society created a very strong social attraction. The emergence of this attraction led to a black hole effect in the Soviet Union, that is to say, during the Soviet period, more national societies joined the Soviet society, and this has formed a larger social-oriented society.

However, after the Soviet society continued to absorb more republics, it meant that many societies that were very different from the social-oriented society also appeared in the Soviet society. The emergence of these affiliated nations and societies with very different cultural heritage directly affected the social interaction in the Soviet society. Possibly even weakened social interaction in Soviet society.

In order to prevent the further increase of the total amount of information and to stabilize the social interaction of the Soviet society, the method of information restriction was adopted within the Soviet Union. This restriction on the free flow of information is mainly reflected in the official media. For example, the Soviet Union's "Pravda" and so on dominated the information exchange of the entire society. On the other hand, it interfered with external information, such as Western radio broadcasts, to prevent Western information from being transmitted to the Soviet Union. These alleviated the information accumulation effect of Soviet society to a certain extent. But while it was stable on the surface, it actually greatly weakened the social interaction within the Soviet society. The weakening of social interaction will of course cause a series of thermodynamic effects. These effects have been analyzed in great detail in previous chapters.

It can be seen from the above analysis that, on the one hand, the information interaction within a social-oriented society will lead to a rapid increase in the amount of information. And this rapid increase in information has a cumulative effect. The more the total amount of information increases, the more powerful it will be to the surrounding society and social individuals. Thereby promoting the expansion of social-oriented society. But on the other hand, due to the increase in the number of external societies and social individuals absorbed by the whole society, it has led to differences within the entire social-oriented society. This intra-society division is an important factor in weakening the interaction within the social-oriented society. In order to resolve differences, some typical social-oriented societies will use coercive methods to limit the transmission of information and limit the accumulation of the total amount of information. And this restriction on the total amount of information directly weakens the social interaction within the social-oriented society. The weakening of social interaction between social individuals is the most important reason for the final collapse of social-oriented society.

7.2.3 Case study: The interaction of social-oriented society from the perspective of classic literary works

From Tolstoy's "War and Peace" and "Anna Karenina", we can see that in a social-oriented society, individuals of different genders pay more attention to the direct and simple interaction between the sexes. The family only plays a role in hindering gender interaction. Like the relationship between Anna and Vronsky, the true love described in the novel is blocked by Anna's family. And ended up ruining both of them.

Even the film "Moscow Does Not Believe in Tears" in the late Soviet Union basically continued such a cognition of gender interaction: a direct and simple effect, just like the interaction between electrons and protons. Among them, the family is the main obstacle to the further development of the gender relationship and the formation of gender molecular. The true love is between Katerina and Gosha. But both parents seem to be non-existent. The development between Katerina and Rachkov that should have continued, was destroyed by the presence of Rachkov's mother.

On the other hand, classic literature in family-oriented society, such as ancient Chinese classics, the works like "Jin Ping Mei" reflect the family's emphasis on the safety of gender interaction through the direct description of sexual relations. Overall, in Chinese literature, family has a positive effect on gender interaction. Of course, since there are positive effects, there are also negative effects. It is clear that bad families are detrimental to safe gender interaction. For example, the Chinese classical literature "Liang Shanbo and Zhu Yingtai" and so on. It can be seen from this that in the family-oriented societies, the mechanism of gender interaction is very complex.

The same is true of classical literature in European family oriented societies. For example, the Italian 14th century novel "Decameron", which also has a very detailed description of sex. This is unimaginable in a social society. Although it is the play "Romeo and Juliet" written by Shakespeare in England, the story takes place in a family-oriented society such as Italy, and we can also see the very important connection between family and gender interaction. Of course, judging from the book "Lady Chatterley's Lover" by the late British writer Lawrence, it is possible to write a work with a slightly more intuitive description of gender interaction, which also reflects that the British period has begun to enter a stage of balanced society. This may be related to the fact that Britain has begun to absorb some of the more typical family-type cultures in the colonies during this period.

7.3 The two states and development of social-oriented society.

7.3.1 Gaseous state

The possibility of social-oriented society becoming a gaseous society is relatively high. This is mainly due to 3 factors.

The first factor is social-oriented society, which is generally located in the colder regions of the north. In these cold regions, the growth rate of vegetation is relatively slow, which causes the social individuals in the social-oriented society to survive mainly in the form of nomads. Therefore, in the cold north, population density is unlikely to increase very much. This makes the social space of the social-oriented society have great potential for expansion.

The great potential of social space expansion means that in the cold northern regions, the distance between basic society members of social-oriented society is relatively far away. This relatively long distance limits the interaction between social individuals in traditional societies. This further weakens the interaction coupling coefficient between social individuals in social-oriented society.

The second factor is that the interaction coupling coefficient of social individuals in a social-oriented society is also affected by the ability of leaders. If there is a strong leader like Touman Chanyu in the social-oriented society. It will greatly enhance the social interaction coupling coefficient. However, the emergence of such powerful leaders was accidental. More often it is a leader who does not have much cohesion. In this case, the social interaction coupling coefficient in the social-oriented society will drop rapidly.

The third factor is also related to the application of information technology in social-oriented society. In the modern social-oriented society, due to the rapid development of information technology, the social space owned by social individuals has increased rapidly. However, in some social-oriented societies, in order to avoid the increase of the total amount of social information, more information control methods are adopted. This also weakens the social interaction coupling coefficient between individual members of society.

It can be seen from the previous analysis that if the social interaction coupling coefficient decreases and the social space of individual members of the society increases, this will rapidly reduce the critical temperature of society. The decrease of the critical temperature is an important sign that the social thermodynamic system cannot enter the condensed state.

7.3.2 Liquid state

Contrary to the gaseous social-oriented society, to realize the social-oriented society also reaches a state of condensation, this time is mainly achieved by raising the critical temperature and reducing the individual social space.

An increase in critical temperature can be achieved by enhancing the interaction between individuals. However, when the interaction between individuals is enhanced, it also means that the information exchange between individuals will be more sufficient. Information technology can enhance the exchange of information between individuals. It can also impair the exchange of information between individuals. Therefore, in improving the interaction and technology between individuals, information technology should be used correctly and appropriately to achieve the goal.

Another aspect is to introduce a strong leader in the social-oriented society. But the problem is that human leadership is random. Once a social-oriented society enters a gaseous state, it becomes very unstable. Just like the Hun society, once the leader dies, the whole society collapses and ceases to exist. Therefore, strengthening the interaction between social individuals by means of strong leaders can improve the stability of social-oriented society in the short term, but it will bring more instability and destructive effects in the long run.

The main aspect of reducing the social space of social individuals is to control the use of information technology. If all individuals in a social-oriented society can be controlled not to use information technology to communicate, the information social space of social individuals will be relatively small, which will help to raise the critical temperature. But such an approach requires the entire social-oriented society to be closed for management. At present, it is difficult to do this in the context of globalization. At the same time, restricting the use of information technology also restricts the development of the whole society. The continuous application and penetration of modern industrial technology will cause the social temperature of this society to be higher than the ambient temperature. This directly results in the inability of this society to effectively obtain energy from the environment. Famine and massive malnutrition followed. This also affects the stability of society.

7.3.3 Development of social-oriented society

After all, social-oriented society is also a result of the long-term development of human history. The formation of social-oriented society is also in line with the requirements of social thermodynamic system. Therefore, compared with other social forms, there is no question of whether a social-oriented society is good or bad.

From the previous analysis, we can already see that the problems affecting the development of social-oriented society are mainly concentrated in the critical temperature being too low. This will eventually make it difficult for social-oriented society to form a condensed state.

If the social-oriented society does not form a condensed state, it means that the society lacks sufficient stability. The interaction of various social thermodynamic parameters will inevitably lead to a wide range of changes in the social space of the entire system.

In the ancient social-oriented society, solving this problem was mainly accomplished by excellent leaders. Once an excellent leader emerges in a social-oriented society at a certain period, a very strong cohesion will be formed within the social-oriented society at this time, and eventually the social-oriented society will form a condensed state.

However, under the current development conditions of globalization, through excellent leaders to enhance the interaction between social individuals, this method can easily lead to various problems. For example, the issue of competition with neighboring countries, various strategic position issues, and so on. On the other hand, what is more important is that when modern technology is developing rapidly, and even when human industrial technology has begun to affect the global environment, it is even more necessary to give full play to the ingenuity of social individuals, so as to be able to design better ways to deal with it, and to promote the stability of the entire human society. On the one hand, strong leaders enhance the interaction between social individuals, but on the other hand, they also suppress individuals' thinking, resulting in the inability of social individuals to think independently, thus affecting the discovery of new scientific laws.

As for the method of reducing the social space of social individuals by restricting the use of information technology by social individuals, it has also been proved to be inappropriate now. Restricting the use of information technology by individuals in society means hindering the technological, cultural and economic development of the entire society. The whole society will be in a relatively backward state. But at the same time, we cannot ignore the impact of industrial technology on human social life. The application of various industrial technologies has led to a rapid rise in the social temperature of society, which is very damaging to traditional societies.

Therefore, to develop social-oriented society, it must conform to the trend of technological development and make extensive use of information technology. The use of information technology allows individual members of society to communicate more freely with each other, rather than suppressing communication between individual members of society. In this way, the interaction coupling coefficient between individual members of society can be quickly improved.

On the other hand, the improvement of the interaction coupling coefficient between individual members of the society can offset the increase of the individual social space. Finally, the critical temperature of the social thermodynamic system can be maintained at a relatively high level.

References

Cui, J., & Chang, H. (2013). The possible climate impact on the collapse of an ancient urban city in Mu Us Desert, China. *Regional Environmental Change*, 13(2), 353–364.

Engerman, D. (1999). Engerman on Firth and Noren, 'Soviet Defense Spending: A history of CIA estimates. H-Net Reviews.

Firth, N. E., & Noren, J. H. (1998). *Soviet defense spending: A history of CIA estimates, 1950–1990* (Vol. 58). Texas A&M University Press.

Fox, C. J. (1976). The politics of culture in the USSR: Art and the Soviet Government. Fletcher F., 1, 204.

Gallagher, D. (2022). Russian Orthodox Church alleges gay pride parades were part of the reason for Ukraine war. https://edition.cnn.com/onfuc/live-news/onfuci-russia-putin-news-03-08-22/h_de0516e0f59ac2214af21bbb0aaf152e

Gunchinsuren, B. (2017). The archaeology of Mongolia's Early States. In *Handbook of East and Southeast Asian Archaeology* (pp. 707–732). Springer.

Kelly, C. (2010). *The end of empire: Attila the Hun & the fall of Rome.* W.W. Norton & Company.

Kim, K., Brenner, C. H., Mair, V. H., Lee, K. H., Kim, J. H., Gelegdorj, E., … & Kim, K. Y. (2010). A western Eurasian male is found in 2000-year-old elite Xiongnu cemetery in Northeast Mongolia. *American Journal of Physical Anthropology*, 142(3), 429–440.

Obrusanszky, B. (2009). Tongwancheng, the city of Southern Huns. *Transoxiana*, 14.

Sandomirsky, V. (1951). Sex in the Soviet Union. *Russian Review*, 10(3), 199–209.

Thompson, E. A., & Drinkwater, J. F. (2015). Huns. OXFORD CLASSICAL DICTIONARY. Retrieved May 23, 2023, from https://oxfordre.com/classics/display/10.1093/acrefore/9780199381135.001.0001/acrefore-9780199381135-e-3175.

Vaissière, E. (2006). XIONGNU. Encyclopædia Iranica, online edition, Nov, 15. Available at https://www.iranicaonline.org/articles/xiongnu. (Accessed on 22 April 2022).

8

Family-oriented society

8.1 The origin and development of family-oriented society

8.1.1 The influence of warm climate on the formation of family-oriented society

If the environment of human social life is very warm, at this time, it can be found from the calculation of case study 6.2.3 that the temperature of the society formed by human beings, even in the traditional social state, can satisfy the social temperature lower than ambient temperature requirements. Under the condition of obtaining energy from the environment very naturally, as long as human societies meet the requirements of interaction required by general social thermodynamics, the whole society can be in a state of condensed.

Unlike the cold weather in the north of the earth, the warm and humid oceanic climate in the south can meet the requirements for plants to grow well in at least three seasons of spring, summer and autumn. In this way, under the condition of sowing in spring, growing in summer and harvesting in autumn, especially the sowing of crops such as rice, can continuously provide sufficient matter energy for the traditional society.

On the other hand, the turmoil in the north at high latitudes prompted a large number of ancient Chinese populations to migrate south. The stable situation in the South is more suitable for the formation of an independent family-oriented society system. For example, in Southeast Asia, the ethnic composition is complex, the number is large, and the language is complex. This is evidence of the long history of the development of independent family-oriented social systems. The single language and simple ethnic structure in the north are the result of a large number of social integrations.

Since human societies can continuously obtain energy in one place, they do not need to migrate as frequently as they do in the north. Over time, those places where certain members of society lived for long periods of time developed a concept of so-called private property. That is to say, these social individuals believe that the assets that they have inherited from their parents or even earlier ancestors must belong to them and cannot be violated by others. This is very different from the social-oriented society. In the nomadic society in the north, as the seasons change, the whole society needs to obtain energy in different places, so the concept of private property is not so strong.

According to the ecological point of view, if a society has no natural enemies, but can easily obtain enough energy from nature, then the population of this society will continue to grow (Seidl & Tisdell, 1999). This growth is exponential until thermodynamic equilibrium is reached.

Since these social individuals all have the same ancestors, these social individuals actually constitute a family relationship. This kind of kinship will make the social interaction between these social individuals stronger. This does not require a very strong leader to achieve the effect of coupling coefficient enhancement, as a social-oriented society does. Even without any one leader, such a society is able to achieve a fully autonomous thermodynamic equilibrium. This is also an important difference between family-oriented society and social-oriented society.

In family-oriented society, the information interaction between families is relatively weak. But this is different from the interaction between different tribes in the cold northern regions. The main reason is that the weather in the south is relatively warm, so it is rich in products and the acquisition of energy is very free. In this way, the population density in the south is very large. Even if there are different families, there is still enough information exchange between families. Therefore, in family-oriented society, although the information interaction between families is weak, it still exists, and with the help of covalent bonds based on gender interactions, the information interaction between multiple families can still form a liquid society.

In the cold regions of the north, due to the vast area and the vegetation of the grasslands, only one or a few tribes can use it. This resulted in basically no mutual information interaction among the tribes in the northern cold regions. Of course, there is also a small amount of information exchange, but the social interaction caused by this information exchange is very weak and can be completely ignored. Therefore, unless a very strong leader emerges, the social-oriented society formed in the cold northern regions is generally in a gaseous state.

8.1.2 Early family-oriented society

According to archaeological findings, the earliest family-oriented society should have originated from the Hemudu society nearly 7,000 years ago or more (Guoping, 2013). In addition, Egypt has a longer history of civilization. It is believed that if there is more archaeological evidence, the development history of Egypt's family-based society may be longer. Such evidence includes the ability to grow crops such as rice and the ability to raise livestock to meet the household's autonomous energy needs. At present, a large number of relics of rice cultivation have been found in the ruins of Hemudu. And this site has continued for thousands of years of history. Existing discoveries have shown that in today's Shi'ao ancient rice field site in Yuyao City, Zhejiang Province, the paddy field area has reached 900,000 square meters (Li, 2021). It can be estimated that the scale of the family-oriented society in that year was very large.

With the cultivation of rice, it means that this society has been able to grow rice to obtain various nutrients without relying on external commercial trade.

Once there is a parent's gender interaction, children are produced. At this time, a relatively solid social interaction within the family based on gender interaction will be formed. In the process of interaction with the surrounding environment, these children will obtain different social divisions of labor to jointly undertake family responsibilities.

If you put aside the gender interaction between parents, in fact, in family society, a family can often be regarded as a very small society, and this small society must be in a very strong internal social interaction.

This can be seen from the discourses of ancient Chinese Confucian philosophers, who actually regarded the family as a miniature version of a society. The so-called "harmonizing the family, governing the state". That is to say, if you can manage your own family well, then you will be able to manage a large society like a country, because each family is a small society.

Due to the complexity of gender interaction, there was a small number of polygamous or polyandrous family structures in early family societies. The most common of these is polygamy. Of course, most families are monogamous.

In terms of scale, the family size in family-oriented society is also very large. For example, a couple has three children, and then the three children continue to multiply, and they will continue to add nine children. Thus the whole family has 14 members. All 14 family members will have very close ties. In the process of social development, these family members will work together throughout their lives to maintain the integrity of the family. The way to maintain this close contact can be through daily communication, parental leadership, family division of labor and cooperation, and annual ritual activities on fixed festivals. Compared with social-oriented society, after a couple has three children, the children will soon become independent. In this way, the whole family will only have five members at most.

Different from the modern family, the early family is gathered in a relatively small physical space. This facilitates collaboration among family members.

Considering that farming has been able to satisfy the self-sufficiency of the entire family members, there is less interaction between one family and other families. However, because the warm climate is more suitable for human habitation, the population density of family-oriented society tends to be relatively large. In this way, the reduction of physical space promotes the mutual communication between different families to a certain extent. However, it does not mean that it is impossible for family-oriented society to form a strong information interaction between basic society members as in social-oriented society. For example, the Hakka people in southern China maintain the stability of the entire Hakka society by inheriting the long-standing Confucian culture.

8.1.3 Case study: Hakka people in southern China

In the family-oriented society, there are also ethnic groups that migrate as widely as the Hun society. This is the so-called Hakka people of southern China (Erbaugh, 1992; Leong, Wright, & Skinner, 1997; Jones, 2010). And the number of people far exceeds the Hun society. According to historical records, the Hakka people in southern China had four large-scale migrations. After the earlier "Eight Kings Rebellion" in the Western-Jin Dynasty, the number of Hakka people who moved south to escape the war reached about two million. In the three subsequent migrations, the numbers increased. There are currently more than 80 million Hakka people around the world. Most of these are descendants of the Hakka people themselves. In contrast to social-oriented society, although the population of the Soviet Union reached more than 200 million. However, the social individual of the Soviet Union was formed by the joining of independent nationalities at high latitudes.

Just different from the Hun social-oriented society, the Hakka people rely on the close interaction within the family, coupled with the promotion of external pressure and the active inheritance of traditional Chinese Confucian culture, which enhances the social interaction between the outside of the family and other families, so the closer ties were formed. Throughout history, the condensed state formed by the Hakka has remained intact for more than a thousand years.

It is precisely because they can maintain a condensed state for a long time that the Hakka people's willingness to expand their social space is relatively weak. On the one hand, they call themselves "Hakka" (it means "guests" in Chinese), that is, they regard themselves as guests from other places, and they can always respect the customs and habits of the local people and live in peace with the local people. Therefore, it is also necessary to restrain as much as possible the desire to expand social space. This is fundamentally different from the colonialism of the European balanced society in the seventeenth and eighteenth centuries.

Of course, when faced with more serious external pressure, the Hakka people will also resist for a short time. It was mainly the uprisings of the Hakka people during the Yuan and Qing dynasties. Among them, the Taiping Heavenly Kingdom even attacked Nanjing, but it was quickly suppressed by the Qing Dynasty. More often, the Hakka people still use their own potential to find more other living spaces for migration.

Chinese Hakka people can be seen as a very typical case of family-oriented society migration. The reason for migration is to maintain the integrity of their own culture. Therefore, at present, the Hakka people are the most complete ethnic group with Chinese traditional culture preserved. A lot of classical Chinese philosophy, art, culture, etc. can find very sufficient evidence from Hakka culture. Even the pronunciation of ancient Chinese can be found in the full chain of evidence in the Hakka language.

From the cultural traditions of the Hakka people, we can see some important characteristics related to the interaction within the family:

One is to emphasize the importance of family education. Since the end of the Western-Jin Dynasty, the Hakka people have attached great importance to the education of their children. The content of education is mainly traditional Confucianism. This education is even gender-neutral. Many Hakka daughters are also well-educated (Ardizzoni, 2021).

The second is to attach importance to serving the country. If the descendants of Hakka can implement effective family education and the whole family is managed in an orderly manner, they can further develop their talents to govern the country.

The third is to emphasize the importance of neighbors. In the "Three Character Classic", there is an allusion: Mencius' mother moved three times in order to allow Mencius to develop good habits, so as to choose a good neighbor and let Mencius get a positive influence. It proves that in family-oriented society, the interaction between families is also highly valued. This is also an important evidence that social interaction in family-oriented society can be enhanced through traditional Chinese culture.

Fourth, it emphasized the importance of respecting the old and loving the young. Not abandoning the elderly and caring for young people and children are important measures to maintain family integrity and family size.

It can be seen from the above characteristics that within the Hakka people, the interaction between families is still relatively strong. However, since they all claim to be "guests", it also shows that the social interaction between the local families in their migration areas is still relatively weak. This is also an important feature of family-oriented society. At this point, the Hakka people still have their own characteristics.

Because the Hakka people have the above-mentioned characteristics that are very different from other ethnic groups in the world. We can also calculate some important thermodynamic parameters of the Hakka society here, and then compare it with the Hun society, there will be some interesting findings.

Considering the current number of Hakka people is very large and distributed all over the world. Therefore, here we take the ancient Hakka people as an example for calculation and analysis. If the number of Hakka people at that time is about 4 million, then

$$N_{hakka} = 4 \times 10^{6}$$

Then the daily nutrition of the Hakka people will be less than that of the Huns. But due to the need to migrate, more energy needs to be added. Therefore, it can still be calculated according to the calorie intake of 1,400 kcal per day. Then

$$dE_{nutrition} = 1400 \times 4184 \times 4 \times 10^{6} \approx 2.33 \times 10^{13} \,(J)$$

Of course, because the migration process of Hakka people needs to consume more extra energy, the extra energy that needs to be consumed will account for a slightly higher proportion of nutritional energy. Take 60% here. Then the social temperature of Hakka society is

$$dT = \frac{dU}{C_v} = \frac{dU}{N\beta} = \frac{1.6 \times 2.33 \times 10^{13}}{4 \times 10^6 \times 2.76 \times 10^4} \approx 336(K)$$

It can be seen that the social temperature of Hakka society is still relatively high. This means that it is difficult for the Hakka people to independently obtain energy from the environment by their own power. This may also be the reason why the Hakka people need to migrate to an area with a higher ambient temperature. That is to say, in addition to historical wars and other reasons, there are actually more reasons for the migration of the Hakka people due to their internal factors. Under such conditions, whether the Hakka society can form a condensed state is crucial. If a condensed state cannot be formed, the Hakka society needs to seek opportunities for expansion on the spot.

Now let's calculate the critical parameters of the Hakka people. First of all, considering that the migration of Hakka people basically involves a large number of old and young women and children, the migration of Hakka society is basically carried out on foot. In this way, the social space occupied by each Hakka social individual can be calculated according to the normal value, that is,

$$v_{hakka} = 0.06m^3$$

Then

$$b = 4 \times 10^6 \times 0.06 = 2.4 \times 10^5 m^3$$

Thus we get the critical social space of the Hakka society as

$$V_C = 3b = 7.2 \times 10^5 m^3$$

Because the Hakka people belong to a family-based society, there are very strong social interactions within the family. And through traditional Chinese culture, social interaction can also be enhanced to a certain extent. Therefore, the coupling coefficient of social interaction within the Hakka society will be stronger than that of the Hun society. Its coupling coefficient can be taken as $a_{hakka} = 1.5 \times 10^{10}\, Pa \cdot m^6$. That is, it is ten times stronger than the social interaction coupling coefficient within a general social-oriented society. Therefore, the critical pressure is:

$$p_C = \frac{a_{hakka} N^2}{27 b^2} = \frac{1.5 \times 10^{10} \times 16 \times 10^{12}}{27 \times 2.4^2 \times 10^{10}} = 1.58 \times 10^{11}\, Pa$$

Substituting into formula (6-38), we can find the critical temperature as

$$T_C = \frac{8 a_{hakka}}{27 \beta v_{hakka}} = \frac{8 \times 1.5 \times 10^{10}}{27 \times 2.76 \times 10^4 \times 0.06} = 2.74 \times 10^6\, (K)$$

It can be seen that this is a very high critical temperature. Therefore, as long as the Hakka people reasonably control the thermodynamic parameters such as social pressure and social space, the entire Hakka society can be kept in a condensed state.

Although at present, the Hakka people are mainly distributed in the mountainous area of about several hundred thousand square kilometers in southern China, the actual area where the Hakka people gather is still very small. To estimate according to the social temperature of Hakka society 336K, and substituting into formula (5-11), we can get:

$$s = \frac{V}{3} = \frac{a_{hakka} N}{3 \beta T} = \frac{1.5 \times 10^{10} \times 4 \times 10^6}{3 \times 2.76 \times 10^4 \times 336} \approx 2.2 \times 10^9\, (m^2)$$

This means that if the social temperature of 336K is to be maintained, four million Hakka people will gather in an area of about 2,200 square kilometers. This is equivalent to the area of a modern city. For the vast southern region of China, it can easily accommodate so many Hakka people.

In this condensed state, we can also calculate its internal pressure:

$$p_i = \frac{a_{hakka}N^2}{V^2} = \frac{1.5\times10^{10}\times16\times10^{12}}{6.6^2\times10^{18}} \approx 5.63\times10^3\,(Pa)$$

Compared with the Hun society under the leadership of Touman Changyu in Case study 7.1.3, this internal pressure is relatively small. However, it is still a relatively large number.

If the Hakka people eventually settle down, avoid conflict with the locals, and gradually disperse, this pressure will be greatly reduced. Social temperatures can also be lowered.

We can also calculate the entropy of the Hakka society in the ancient.

$$S = N\beta\left\{\frac{3}{2}lnT + \ln\left(\frac{V}{Nn_p}\right) + \frac{3}{2}\left[\frac{5}{3} + \ln\left(\frac{2\pi mk_B}{h^2}\right)\right]\right\}$$
$$= 4\times10^6\times2.76\times10^4\times\left\{\frac{3}{2}ln336 + ln\frac{6.61\times10^9}{4\times10^6\times2\times10^{27}} + \frac{3}{2}\times\left[\frac{5}{3} + ln\left(\frac{2\pi mk_B}{h^2}\right)\right]\right\}$$
$$\approx 2.45\times10^{12}\left(J\cdot K^{-1}\right)$$

From the calculation results, compared to case study 7.1.3, the entropy of the ancient Hakka society is larger than that of the Hun society under the leadership of Touman Changyu. This is because the Hakka society has a much larger population. However, since entropy is a variable that can be accumulated. So we can also average this entropy to 1 million people. Then the entropy of each million Hakka society is

$$S \approx 6.13\times10^{11}\left(J\cdot K^{-1}\right)$$

That's smaller than the 1 million Huns in Case study 7.1.3. It shows that compared with the Huns, the Hakka society is more orderly.

8.1.4 Modern family-oriented society

In some aspects, the modern family-oriented society and its structure are similar to the ancient family-oriented society. What sustains the family is the gender interaction of the parents and the social interaction based on the kinship between the children and the parents.

But in other aspects, the modern family-oriented society is very different from the traditional ancient family-oriented society. The family structure of modern family-oriented society is more complex. And through the use of information technology, the members of these families can be distributed in different geographical locations. This can be seen as a distributed family structure.

Despite the development of modern technology, the traditional family seems to have a tendency to be destroyed on the surface. But judging from the performance of the whole society and culture, it can be found that in fact this is only a superficial change. The genes of family culture seem to be very strong throughout the family-oriented society. For example, in terms of corporate culture, companies in traditional East Asian societies are still dominated by family-owned businesses. After a successful business, the most common thing is to let your children inherit it. In European and American balanced societies, this kind of corporate culture is relatively rare.

Although the destruction of this family structure may be more of a superficial phenomenon, it has also attracted the attention of many local governments in East Asia. For example, in order to maintain the traditional family structure in Singapore, various administrative measures have been adopted to support families (Wang, 2020). These include an emphasis on the family as the foundation of society, which encourage family priority and home ownership and introduce support policies to maintain family stability and so on.

Due to the large-scale production of industrialization, this family relationship has even affected the organizational structure of the entire society in modern times, including the internal cultural organization of some large multinational corporations (Kumagai & Keyser, 1996). One fact is that Japanese family structures are generally much more complex than those in the balanced or social-oriented societies of Europe,

America and the Soviet Union (Kitaoji, 1971). Even during the Allied occupation of Japan, the family structure of this family-oriented society was seen as the traditional Japanese family structure (ie) was a product of feudalism and attempted to undermine it at the level of state policy. However, historical facts over the past few decades have shown that such attempts are of little use. Aside from the absence of ie in the name, this form of family structure remains deeply rooted in all aspects of Japanese society (Moon, 2002).

The influence of this family business culture may even be difficult for many Western employees to accept. It has also hindered the expansion of some Asian multinational companies in Europe and the United States. Of course, many Asian multinational companies have already noticed this problem and vigorously integrated into the culture of balanced society in Europe and the United States. As a result, the operating conditions of enterprises have also been greatly improved, and their competitiveness has also been significantly improved.

8.1.5 Case study: Thermodynamic parameters within a family

For a traditional family-oriented society, a family usually consists of such members: parents, children, grandchildren. There are also grandparents from both parents. This relatively small family will reach a dozen members. For example, $N = 18$.

Considering that the gender interaction of parents and grandparents is very strong, there are at least three gender nucleus. This way we can represent the family with a Feynman diagram. As shown in Figure 8-1.

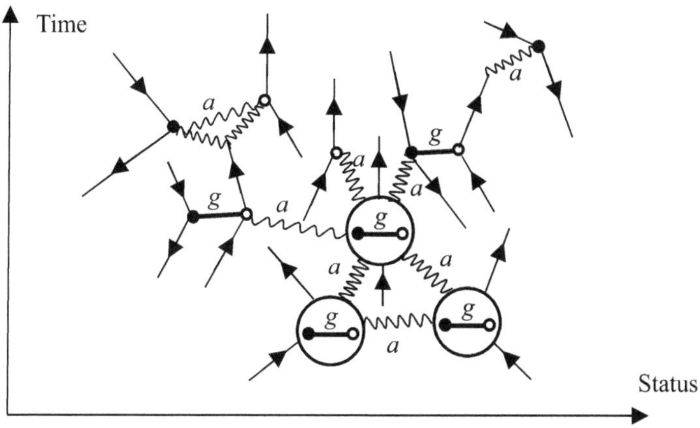

Figure 8.1. The Feynman diagram of a family

After subtracting three gender nucleus, the number of basic society members of the entire family's social thermodynamic system is 15, so $N = 15$.

Considering that the interaction within the family is very large, we can assume that the social interaction coupling coefficient within the family's social thermodynamic system can reach 100 times that of the normal social interaction outside. In this way, the social interaction coupling coefficient within the family can reach

$$a_H = 1.5 \times 10^{11} \left(Pa \cdot m^6\right)$$

Without the use of information technology, the critical temperature of the entire social thermodynamic system will reach

$$T_C = \frac{8a_H}{27\beta v_p} = \frac{8 \times 1.5 \times 10^{11}}{27 \times 2.76 \times 10^4 \times 0.06} = 2.68 \times 10^7 \,(K)$$

It can be seen that this critical temperature is very high. This is also the basic reason why a family can always be in a condensed state.

In addition, if we think that the temperature of the family is the same as the temperature of the society, we can also calculate the social space where the inflection point E of the Van der Waals isotherm shown in Figure 6-3 is located by formula (6-44) as

$$V \approx \frac{2a_H N}{\beta T} = \frac{2 \times 1.5 \times 10^{11} \times 15}{2.76 \times 10^4 \times 272} \approx 6 \times 10^5 m^3$$

It can be seen that the social space where the inflection point E is located is actually very large, which can be converted into an area of 0.2 square kilometers. It can be expected that at such a high critical temperature, the social space at point C will be very large. This means that even if family members are distributed in a very broad geographical location, reaching an area of several square kilometers, they can always maintain the social thermodynamic system of the entire family in a condensed state.

However, we can also see from the above calculation that if the social interaction within a family's social thermodynamic system only maintains a normal level, it means that the whole family's social interaction can only be maintained within a small geographical range, in which the thermodynamic system is in a condensed state. This is the reason why the cohesion within the family is not as good as in the social-oriented society.

8.2 Interaction

8.2.1 Interaction within the family

The internal interaction of family-oriented society is very complex. This complexity is mainly reflected in the fact that in addition to gender interaction within a family, more importantly, in family-oriented society, there are actually very complex social interactions within the family.

In family-oriented society, the interaction within the family mainly has the following characteristics:

First, in a family-oriented society, gender interaction must be ensured, and this gender interaction is an inevitable requirement for the existence of parents. In a family-oriented society, the presence of parents in the family is necessary. If there are no parents, then the family will have no dominant social members. Of course, such a family also degenerates into a general society. There is not much difference compared to the interaction between members of a social society.

Second, in family-oriented society, the intensity of social interaction within the family is very strong. The main reason is that the social interaction within the family is actually an extension of the gender interaction between parents. Because gender interaction is very strong, so this social interaction within the family will also become very strong. It's like two thermodynamic systems in close contact, if the temperature difference between the two thermodynamic systems is very large, the hotter system will transfer heat to the colder system. And if the thermodynamic system with high temperature can continuously generate heat, and the outside of the thermodynamic system with low temperature is a relatively low ambient temperature, a temperature gradient can be formed. The gender interaction of parents can be seen as a thermodynamic system with a relatively high temperature, while the social interaction of children can be regarded as a thermodynamic system with a relatively low temperature. If family members are closely linked, it means that the strength of gender interaction is gradually passed on to social interaction. From this point of view, social interaction and gender interaction are still different from gravitational interaction and electromagnetic interaction in the physical world. After all, the difference between gender interaction and social interaction is not as uncoordinated and ununited as electromagnetic interaction and gravitational interaction.

The third is the complexity of interaction within the family. Due to the existence of this interaction intensity gradient, it also means that there is a process from strong to weak interaction within a family. In a family, there are obvious differences in the intensity of interaction between parents, children, and relatives, so there is a similar quantization effect in the interaction within the family. First of all, the intensity of gender interaction of parents is the largest. This is the core level. Then the children are extended by the parents, which is the second level. In the second level, the intensity of social interaction between children is weaker than that of parents' gender interaction, but it is stronger than that of social members with kinship in the periphery. At the third level are the children of siblings who belong to their parents, with whom they are related. Therefore, there is still relatively strong social interaction between family members with kinship and family members at the second level, but the strength of social interaction at this level is obviously weaker than that between parents' biological children. From this we can see that there is a jump process in the interaction intensity at different levels. It is as if in an atomic system, the change of energy changes by leaps and bounds. It can also be seen from this that a family in a family-oriented society is actually more similar to an atom in the physical world. As shown in Figure 8-2. In social-oriented society, the atom structure is much simpler. Because the social individual in each social-oriented society can be regarded as an atom individually. The gender interaction is directly used to form gender molecules between two different genders.

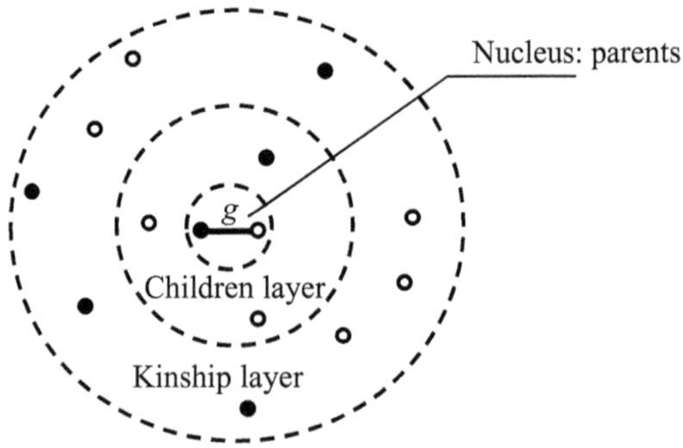

Figure 8.2. The family atom

It can be seen from this that in family-oriented society, a family is more similar to a heavy atom in the physical world, which can form a matter with a larger specific gravity. The social-oriented society is more similar to the state of hydrogen. Of course, this is a static observation perspective, and it is still a bit abstract to intuitively see the interactions between family individuals. The dynamic perspective is still analyzed with Feynman diagrams, as shown in Figure 8-1. This can also show the benefits of Feynman diagrams in describing interactions between members of a society.

The fourth is the privacy of gender interaction. On the one hand, gender interaction plays a central role in family-oriented society, so it is very important. That is to say, gender interaction is highly valued in family society. Within a family, however, the social interaction between all members is also very strong. This means that in the process of completing gender interaction, parents also need to complete social interaction with their children and other family members. Therefore, there are conflicting issues between social interaction and gender interaction. The solution to this problem is for parents to encapsulate gender interaction. As in family interaction, this gender interaction is just a normal existence in a society. Therefore, this gender interaction has very strong privacy in family-oriented society. This can also be seen from the classic literary works of ancient China. In these classic literary works, either gender interaction is not involved at all, or when involving gender interaction, it will form a very direct description. For example, in "Romance of the Three Kingdoms", "Water Margin" and "Journey to the West", the content of gender interaction is basically not involved. However, in classic literary works such as "Jin Ping Mei", there is a very direct description of gender interaction.

The fifth is the responsibility of gender interaction. Since gender interaction plays a central role in the entire family-oriented society, it also means that gender interaction assumes the core content of maintaining the stability of the entire family-oriented society. Therefore, in family-oriented society, gender interaction is very important. For example, in traditional family-oriented societies in East Asia, there are very grand ceremonies for marriage between the sexes. In the world's earliest collection of poems, The Book of Songs, the first poem describes the process of a boy and a girl falling in love and getting married. The poem describes that the boy and girl were in love with each other at first and were willing to marry. But if the boy want to get married, he need various ceremonies at this time, including "Make friends with the girl through music in the first", "Then playing gongs and drums makes her happy" and so on. Of course, this responsibility is in contradiction with the essence of gender interaction, which is also the root cause of many tragedies in the development of human history. For example, "Liang Shanbo and Zhu Yingtai" in Chinese classical literature. Zhu Yingtai's parents want to fulfill the responsibility of gender interaction, and then arrange for Zhu Yingtai to marry a powerful family. But what Zhu Yingtai likes is Liang Shanbo, who has knowledge but not much wealth. Such contradictions eventually lead to both Zhu Yingtai and Liang Shanbo's martyrdom. Of course, the same problem exists in family-oriented societies in Europe, such as the story of "Romeo and Juliet" and so on. Of course, this in turn proves that, at least in the Middle Ages, Italy and China were very close in terms of family culture.

8.2.2 Social interaction

In the family-oriented society, a family is equivalent to an atom in the physical world. Since family atoms are not like physical atoms, each electron has only two states, and each atom orbital can only accommodate a limited number of electrons according to the Pauli exclusion principle. Once the outer orbit of the physical atom is filled with all electrons, then the physical atom has a strong stability, which is the molecular or atom of the noble gas. In contrast, the children of a family atom are highly mobile. Moreover, these children must also be combined with other relative genders within the external family to form a new family. Then this marriage relationship between children also forms a covalent bond between family atoms. Like the covalent bond between physical atoms, this covalent bond between family atoms is also very strong. Therefore, in a family-oriented society, the interaction between families is also relatively strong.

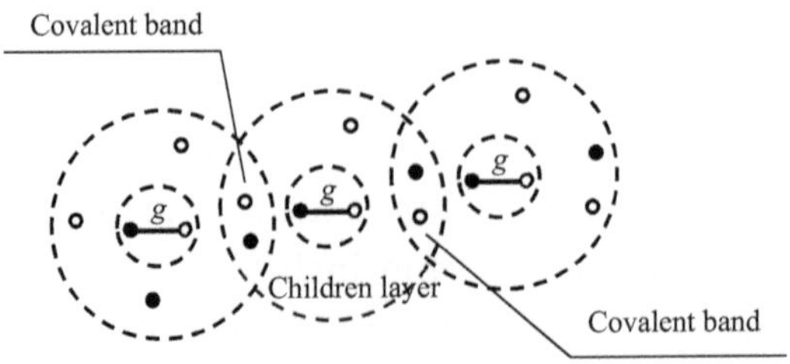

Figure 8.3. The family covalent band

It can be seen from Figure 8-3 that the covalent bond formed by two or more families through marriage is the same as the covalent bond in insulating substances.

Like physical atoms, family atoms can also be combined into an "ionic bond". That is, as children of one family, they can help another family complete some important domestic service work through social division of labor and cooperation. For example, a family needs to install an air conditioner. In a traditional family-oriented society, this work is usually done by the young children of the family. However, in the modern industrialized society, through the division of labor and cooperation, this work can be done by young children of other families. Although the parents of the family do not know which family the young man belongs to, this form of social division of labor can also connect different families together. This can also significantly improve the social interaction between basic society members in modern industrial society.

The early family society mainly relied on a very small number of craftsmen existing in the society due to the relatively small division of labor in the society. But these craftsmen are in a dissociated state in the whole family society, just like free electrons in matter, they can also realize social interaction. However, the social interaction achieved by the social division of labor completed by a small number of craftsmen is still relatively weak. Only in the modern industrial society, the realization of almost all the work requires professionals to complete the strong combination of "ionic bond" between family atoms.

Of course, compared with the family atom's covalent bond, the connection effect of the "ionic bond" caused by the difference in social division of labor is still relatively weak.

In addition, family-oriented social interactions can also refer to cell growth and connectivity models. In this model, the parents are DNA carriers. Through the guidance of parental DNA, family members continue to expand, and family cells gradually improve. After the perfection of family cells, division of family cells occurs. After the children go out independently, they form independent family cells.

The social interaction can be seen as the interaction between cells. This social interaction also directly leads to the close connection of family cells. Forms a solid state similar to the various organs of the human body.

If the connections between such family cells are weak. Then a fluid state like blood can be formed.

The advantage of using this cell model is that there are biological organisms to make an analogy with. Therefore, it is possible to infer changes in the entire society from changes in the physical condition of each individual. This is often used in some ancient philosophical thought. In ancient Chinese Confucianism, it was speculated that the entire social structure was first to deal with personal development, the so-called "checking things for knowledge", and then dealing with family problems, and then dealing with social problems after success.

Of course, just like the various organs composed of organic cells, these organisms need to be kept intact under relatively harsh conditions. More complex biological organisms even need to maintain their body temperature at more than 30 degrees Celsius to remain intact. Therefore, the corresponding family-oriented society also needs to exist under relatively harsh conditions. Once the environment changes, without such conditions, the entire family-oriented society may face serious problems.

Of course, there are still some problems with using this cell model to explain family structure. The main problem is that the interaction between cells is different from the interaction between gas molecules. Once the connections between cells are formed, they are very strong. Rather than a simple function that varies with distance. The interaction between gas molecules can be described by the constant a in the Van der Waals equation.

In addition to realizing social interaction through covalent bond and ionic bond, family-oriented society can also realize social interaction through information interaction. However, because the information interaction within the family is much stronger than the information interaction outside the family, the information interaction of the entire family-oriented society is weaker than that of the social-oriented society.

Considering that although the information interaction in the family-oriented society is very weak, the covalent bond and the ionic bond form a stronger social interaction, so the information interaction between the basic society members in the family-oriented society can be ignored.

However, the social interaction formed by covalent bond or ionic bond also has a disadvantage, that is, the social interaction formed by this covalent bond or ionic bond will become very weak when the distance is relatively long. Therefore, when the basic society members are very far apart, or the population density is very small, information interaction plays a leading role.

Therefore, when the population density is relatively small, the cohesion of social-oriented society will be much better than that of family-oriented society, and the critical temperature of social-oriented society will be much higher.

8.2.3 Calculation of thermodynamic parameters

According to the previous convention on the size of social interactions, that is, if one interaction is stronger than the other, it can generally be estimated by an order of magnitude increase. Precisely because this family-social interaction is more similar to the interaction between large molecules of organic matter, proving that it is stronger than general social interaction. So we can choose here that its Van der Waals constant is about 10 times normal. Which is

$$a_{Hs} = 10a_s = 1.5 \times 10^{10} \left(Pa \cdot m^6\right)$$

Whether it adopts the cell model or the family atom model, it means that the basic society member of family-oriented society is the family. Therefore, the number of basic society members in a family-oriented society is much lower than that in a social-oriented society.

For example, according to the "family atom" shown in Figure 8-2, each family has about 15 members. Therefore, for a family-oriented society with a population of 100 million, the number of basic society members is only 6.7 million. And if it is some more traditional family-oriented society, its family size will be larger, which can reach dozens or even hundreds of people. At this time, the number of basic society members will be reduced on a larger scale.

This will lead to changes in various critical parameters due to increased social interaction a_{Hs}between households. However, since some of these constants also change with the number of family members, this will affect the calculation of critical parameters.

For critical social space, its calculation formula is:

$$V_C = 3b = 3Mv_H \tag{8-1}$$

where v_H is the social space owned by a single family, and M is the total number of families. As the number of family members n increases, the social space owned by a single family is actually the sum of the n family members. So the above formula becomes

$$V_C = 3Mnv_p = 3Nv_p \tag{8-2}$$

Here v_p is the social space owned by a single social individual, and N is the total number of social individuals in this society. It can be seen that the formula calculation of critical volume can still be calculated according to the unit of a single social individual.

The formula for calculating critical pressure is:

$$p_C = \frac{a_{Hs}M^2}{27b^2} = \frac{a_{Hs}}{27{v_H}^2} = \frac{a_{Hs}}{27n^2{v_p}^2} \tag{8-3}$$

In which, n is the number of family members.

It can be seen that the critical pressure is the social interaction coupling coefficient between households divided by the square of the number of household members. Therefore, although the interaction coupling coefficient is increased, the critical pressure is significantly reduced.

In addition

$$\beta_H = n_H k_B = nn_p k_B \tag{8-4}$$

where n_H is the total number of water molecules owned by all members of the family, n_p is the total number of water molecules owned by a single social individual, and n is the total number of family members.

The formula for calculating the critical temperature:

$$T_C = \frac{8a_{Hs}}{27\beta_H v_H} = \frac{8a_{Hs}}{27\beta n^2 v_p} \tag{8-5}$$

It can be seen that compared with the social-oriented society of the same population, the ratio of the critical temperature drop in a family-oriented society is

$$\frac{T_{HC}}{T_{sC}} = \frac{a_{Hs}}{a_s n^2} \tag{8-6}$$

It can be seen that, for family-oriented society, on the one hand, a very strong interaction within the family can easily form a condensed state. This state is very stable.

But we can also notice that when the number of social members in a family increases, the number of covalent bonds between families will also increase, so that the coupling coefficient of social interaction between families will also increase. However, this increase should be an increase in the level of arithmetic, and the square of the number of family members increases faster. Therefore, the conclusion drawn from the above formula remains unchanged.

But outside the home, with the increase of the number of family members, the critical temperature will decrease to a certain extent. This makes it seem more difficult for family-oriented society to form a condensed state than for social-oriented society.

This may be used to explain why in some long-standing traditional family-oriented societies, some strong leaders are also prone to emerge. After all, according to the development history of social-oriented society, the emergence of strong leaders will help to strengthen the social interaction between families. Thereby helping family-oriented society to raise the critical temperature. But given the complexity of this social interaction, not all leaders are able to handle this kind of social interaction between families with ease. Therefore, some measures sometimes adopted may actually lead to the weakening of this social interaction.

For example, suppressing gender interaction, which will lead to the formation of covalent bonds between family atoms is affected. Suppressing freedom of speech will weaken information interaction. These are typical cases of self-defeating, instead of enhancing social interaction between families, it has the opposite effect.

Then we look at the Van der Waals equation:

$$N\beta T_H = \left(p + \frac{a_{Hs}M^2}{V^2}\right)(V - b) \tag{8-7}$$

If the external pressure is relatively small, that is

$$p \ll \frac{a_{Hs}M^2}{V^2} \tag{8-8}$$

can be calculated

$$\frac{T_H}{T_s} \approx \frac{a_{Hs}M^2}{a_s N^2} = \frac{a_{Hs}}{a_s n^2} \tag{8-9}$$

Then

$$T_H \approx \frac{a_{Hs}M^2}{V^2}\frac{(V - b)}{N\beta} \tag{8-10}$$

Therefore, compared with the social-oriented society of the same population, although the critical temperature has changed, the ratio between the social temperature of the family-oriented society and the temperature of the social-oriented society is basically the same as the magnitude of the critical temperature drop. That is to say, the decrease of the critical temperature has no substantial effect on the family-oriented society entering the condensed state when the external pressure is relatively small.

But when the external pressure is large enough, i.e.

$$p \gg \frac{a_{Hs} M^2}{V^2} \tag{8-11}$$

Then

$$\frac{T_H}{T_s} \approx 1 \tag{8-12}$$

Then

$$T_H \approx p \frac{(V - b)}{N\beta} \tag{8-13}$$

It can be seen that the social temperature of the family-oriented society at this time is basically the same as that of the social-oriented society and social temperature of the same population. But the critical temperature of family-oriented society has dropped significantly. This means that family society will be less likely to enter a condensed state and become more unstable. That is to say, family-oriented society is relatively weak against pressure.

Historically, some huge family-oriented societies with a population of several hundred million are facing enormous social pressure amid internal and external troubles. At this time, it is often easily conquered by tens of thousands of military soldiers from social-oriented society or balanced society, which also proves that such instability does exist in family-oriented society.

Of course, this means that the whole society is not easy to enter into a condensed state. But as the previous calculations show, each family is still in a very stable state of cohesion. This is where family-oriented society is very different from social-oriented society. As long as this family structure is not destroyed, when the external pressure is reduced, the entire family-oriented society can quickly enter a condensed state. This can also be used to illustrate that after some major powers successfully occupied family-oriented societies such as Vietnam and Afghanistan, they will face very serious governance problems. Especially over time, after the external pressure disappears, these family atoms can be reassembled to form a condensed state of family-oriented society. It becomes very difficult for external cultures to integrate into these family-oriented societies.

8.2.4 Case study: Urban density comparison of Tokyo and London

As two different social forms, Tokyo in Japan and London in England are more comparable. Japan is a traditional family-oriented society. The UK is a balanced society. Therefore, considering that the number of basic society members in Japan is actually not large, the total number of family atoms is smaller than the total population of the United Kingdom. This way, even if more people gather together, it won't feel too crowded in Japan. Therefore, the Japanese prefer to concentrate in big cities like Tokyo. In the UK, as a balanced society, the difference between the number of basic society members and the population is not as big as that in Tokyo. In this way, even if there are a few million people, each of them may feel the same degree of crowding as a city of tens of millions in Japan.

Let's look at the data for Tokyo first. Tokyo, Japan is the most populous city in the world, with a population of 37.34 million in 2021 (Tokyo Population, 2021) and an area of 2,155 square kilometers.

Correspondingly, in mid-2020, the total population of London is about 9 million (London's Population, 2021), with an area of about 1,577 square kilometers.

According to the calculation of the above data, if only the total population is considered, the population density of Tokyo is 17,300 people per square kilometer. The population density of London is 5,700 people per square kilometer. It can be seen from this that the population density of Tokyo, Japan is far higher than that of London. However, in the daily life of the citizens of the two cities, Tokyo does not actually feel that Tokyo is overcrowded than London.

If we consider that Japan is a family-oriented society, and assume that the total population of each family is 9, the basic society member density is $1,900km^{-2}$. It can be seen that this is lower than the population density of London, UK.

However, if you consider that the UK is a balanced society, it will still contain a certain number of basic family members. Generally speaking, the married population accounts for about 50% of the total population. Consider two married people and their children as members of a basic society, with an average of 4 people in each family. If we think about millions of singles, it can be estimated that the total number of base members in London is about 3.35 million. The density of basic society members calculated in this way is 2,100 per square kilometer. It can be seen that it is slightly larger than Tokyo, but the difference is not as large as the population density.

Therefore, the density of basic society members can more accurately reflect the degree of crowding in a city than the population density. This is because the "crowded" feeling between people is mainly due to the strength of social interaction. If social interaction is weak, even a few people will feel noisy and crowded. But if the social interaction in the group is strong, even if many people gather together, it will not feel crowded. While family-oriented society strengthens the interaction between family members, it also strengthens the interaction between basic society members, which is why living in a city like Tokyo does not have the problem of being too crowded to survive.

Now we can use the above formula to calculate some thermodynamic parameters for Tokyo and London. Because their information technology development levels are similar, and the application of information technology is not necessarily balanced, for example, children and the elderly have relatively weak ability to use information technology, so the social space occupied by a single social individual is set to $v_p = 0.5m^3$

In this way, for Tokyo, the population of each household in Tokyo is calculated according to 9 people, and its critical temperature is:

$$T_{Ctokyo} = \frac{8a_{Hs}}{27\beta n^2 v_p} = \frac{8 \times 1.5 \times 10^{10}}{27 \times 2.76 \times 10^4 \times 9^2 \times 0.5} = 4080(K)$$

The critical temperature is still relatively high. If there is no external social pressure, the total number of households in Tokyo is calculated at 4 million and the area is calculated at 2,155 square kilometers, then its social temperature is:

$$T_{Htokyo} \approx \frac{a_{Hs}M^2}{N\beta V} = \frac{1.5\times10^{10}\times4^2\times10^{12}}{3.7\times10^7\times2.76\times10^4\times2155\times10^6\times3} \approx 37(K)$$

It can be seen that the social temperature is significantly lower than the critical temperature. The whole society is relatively stable.

You can also calculate its internal pressure:

$$p_i \approx \frac{a_{Hs}M^2}{V^2} \approx 5890(Pa)$$

It can be seen that the internal pressure is not large.

Then we calculate the thermodynamic parameters for London.

Although a London household may have an average of 4 individuals. However, since London is a balanced society, in fact, if factors such as independent social individuals and gender elements are taken into account, each basic social member contains about 2–3 social individuals. In this way, the population of each basic social member in London is calculated according to 3 people, and its critical temperature is:

$$T_{Clondon} = \frac{8a_{Hs}}{27\beta n^2 v_p} = \frac{8\times1.5\times10^{10}}{27\times2.76\times10^4\times3^2\times0.5} = 36700(K)$$

It can be seen that this critical temperature is significantly higher than Tokyo. This is due to the smaller number of people per household in London.

We can approximately estimate London $M \approx 3\times10^6$. Then the total population is 9 million, and the area is calculated according to 1,577 square kilometers, then its social temperature is:

$$T_{Hlondon} \approx \frac{a_{Hs}M^2}{N\beta V} = \frac{1.5\times10^{10}\times3^2\times10^{12}}{9\times10^6\times2.76\times10^4\times1577\times10^6\times3} \approx 118(K)$$

Of course, the social temperature is significantly lower than the critical temperature, which also proves that the social structure of London is still very stable. However, this social temperature is already significantly higher than that of Tokyo.

You can also calculate its internal pressure:

$$p_i \approx \frac{a_{Hs} M^2}{V^2} \approx 6180(Pa)$$

This internal pressure is greater than that in Tokyo. But the difference is not huge.

Now consider if the two cities are facing a larger external pressure for some reason. This external pressure is much greater than the internal pressure. Here, set $p = 10^6 Pa$. The calculation formula of social temperature should be calculated by formula (8-13).

Then for Tokyo, its social temperature will rise to

$$T_{Htokyo} \approx \frac{pV}{N\beta} \approx \frac{10^6 \times 2155 \times 10^6 \times 3}{3.7 \times 10^7 \times 2.76 \times 10^4} \approx 6330(K)$$

It can be seen that the social temperature in Tokyo is very high, which has exceeded the critical social temperature of Tokyo. That is to say, in this case, Tokyo's society can only be gaseous.

London's social temperature will rise to:

$$T_{Hlondon} \approx \frac{pV}{N\beta} \approx \frac{10^6 \times 1577 \times 10^6 \times 3}{9 \times 10^6 \times 2.76 \times 10^4} \approx 19000(K)$$

Although the social temperature in London seems to be higher. But we can note that the social temperature in London has not exceeded the critical temperature. That is to say, under the conditions of effective organization in this situation, from the Van der Waals equation, there is still a chance to enter a condensed state by reducing the social space.

Therefore, it can also be found from the calculations that although the critical temperature in Tokyo has dropped due to the change in the number of basic society members, the reduction in the social temperature can solve the problem that the entire society is in a condensed state. Therefore, as long as the external social pressure is not too great, the feeling of living in Tokyo, Japan at this time is actually similar to that of living in London, England.

Of course, when faced with greater external social pressure, Tokyo will be more unstable than London.

8.3 Three states of family-oriented society

8.3.1 Gaseous family-oriented society

The formation of gaseous family-oriented society is mainly caused by the large external pressure. Such external pressures may be foreign invasions, colonial rule, and social changes. From the solution results of Section 8.2.3, it can be seen that even if the social interaction of family-oriented society increases, the critical temperature will still decrease. If external pressure is weak, the social temperature of family-oriented society will also drop synchronously. Such a drop in critical temperature does not produce significant gaseous effects. But if the external pressure is too great, the family-oriented society will have the same social temperature as the social-oriented society with the same number of social individuals. However, the critical temperature of this family-oriented society has dropped significantly, making it difficult for this family-oriented society to enter the condensed state.

Of course another reason for the transformation of family-oriented society into a gaseous state is the very weak social interaction between families. From formula (8-5), we can see that if the social interaction between families is relatively weak, the critical temperature of family-oriented society will be lower than that of social-oriented society of the same population. This situation still occurs frequently in human society. For example, the implementation of the apartheid policy has resulted in the restriction of measures to promote social interaction such as information exchange and intermarriage between different races, thereby weakening the social interaction between families.

In addition, the caste system like traditional Indian is also a factor that weakens the social interaction between families. Intermarriage between different castes is strictly restricted, resulting in the inability of a large number of families to be closely united through intermarriage. The social interaction coupling coefficient between families is naturally very low. In addition, the family size of this kind of traditional society with a long history is relatively large, so under the condition of relatively large external pressure, such a family-oriented society will basically be in an unstable state. For example, in the 18th century, the British army of tens of thousands of people could defeat India, which has a population of nearly 100 million. Fortunately, India's natural environment is very advantageous. The suitable climate allows different ethnic groups and societies in India to obtain enough energy under the unstable conditions of the society as a whole, and maintain a long historical heritage.

Compared with the liquid or solid family society, the family atom in the gaseous family-oriented society has certain fluidity and expansion. That is, because of changes in the external environment and social changes, the entire family atom will flow to a place that is more suitable for obtaining energy. For example, in the process of urbanization in India, a large number of peasant families who lost their land began to flow to the cities and gather around the cities to form so-called "slums". This is also the case in some countries in South America.

Of course, when there is a problem with the foundation of the family's existence, it will also cause the family atom to become a gaseous atom that can move freely. The whole society is in a gaseous state, and it is easy to alleviate the increase in the internal temperature of society by expanding social space like social-oriented society. For example, in the Sanyasi Rebellion (Kundra, 2013) that broke out in Bengal between 1763 and 1800, a large number of farmers lost their land due to colonial rule, so that subsistence farmers could not normally obtain energy from nature. In order to lower the social temperature, landless peasants began to expand their social space through uprisings.

8.3.2 Liquid family-oriented society

Liquid family-oriented society is mainly those social forms that are relatively mobile. This fluid family-oriented society is somewhat similar to the northern nomads in form. But the two types of society are different.

A typical liquid family-oriented society is the Roma in Europe (McGarry, 2014; Cahn & Guild, 2010) and so on. In addition, we mentioned earlier that the Hakka people in southern China had experienced many large-scale migrations, but the mobility of the Hakka people in ancient China was not so strong. Because the movement of these Hakka people only occurs on a large scale when there is war. Once the situation stabilizes, the Hakka usually develop in an area for a long time. The Roma, on the other hand, migrated across the entire European continent.

The main reason for the migration of nomads is the change of grassland. If you graze in one area for a long time, the flock can easily run out of food. In this way, after the food in one area is eaten up, it is the best choice to move to another place. The migration of the Hakka people is mainly caused by social pressure. When the war in northern China made the traditional family-oriented society unable to survive, a large number of Hakka people had to migrate to the south.

Roma migration, however, does not seem to have much to do with these factors. From a physical thermodynamic point of view, one cause of fluid flow is primarily a change in potential energy. When a region's potential energy is high, fluid flows from that region to regions with lower potential energy. In addition, if it is subjected to an external force, it will also cause the fluid to flow in the direction of the force. Considering the generally poor economic conditions of the Roma, some customs are somewhat incompatible with the local mainstream ethnic groups. At the same time, the Roma themselves have some artistic specialties, so the movement of the Roma may be mainly related to social pressure. After living in one area for a period of time, because of the inability to integrate into the local mainstream society, there is a conflict with the local people. At this time, it is easy to be excluded by the local people, which in turn causes the Roma to start to migrate to another place.

From the Van der Waals isotherm, it can be seen that the liquid society occurs mainly on the curve below the critical temperature. If the social space is too small, a society will appear in the condensed state. In a condensed state, although external pressure can vary greatly, the impact on social space is small. The same is true of liquid family-oriented society. Like the Roma in Europe, although they have been living on the fringes of society for more than a thousand years, their social space is very small, and they have also been excluded from the mainstream society. But the whole Roma society can always remain stable and carry on its own culture.

Compared with the liquid social-oriented society, it is more similar to the heavy atom in the physical world due to the larger number of members that make up the family. This results in a more stable liquid family-oriented society.

8.3.3 Case study: Roma society in Europe

The Roma are the most populous minority in Europe, with a population of around 10 million. Roma seem to find it difficult to integrate into European society. Due to less communication with other ethnic groups, it is difficult for them to be understood and accepted by mainstream European ethnic groups, and discrimination in all aspects of life (Cahn & Guild, 2010). But on the other hand, we can also notice that in some regions with strong traditional family culture, such as other parts of Asia, there was also an influx of Roma. The Roma were able to integrate into the cultures of these regions so smoothly that there is little information about the Asian Roma (Gypsies) (Sinclair, 1907).

The Roma originate from a traditional family-oriented society in northwestern India (Mendizabal et al., 2011). Therefore, compared with other ethnic groups in Europe, the Roma have the characteristics of a very typical family-oriented society. From the perspective of Roma societies scattered around the world, Roma usually have less intermarriage with foreigners. This also means that the Roma culture can always be passed on among their ethnic groups, which in turn makes the ties between Roma families very strong.

We can estimate critical thermodynamic parameters and social temperature for Roma.

First of all, we consider that the life of the Roma has always remained consistent with a very traditional family society. Social interaction and individual social space can always maintain their original values for a thousand years. Thus we can calculate

$$a_{Hs} = 10a_s = 1.5 \times 10^{10} \left(Pa \cdot m^6 \right)$$

If the size of each Roma family is about $n = 15$, then for the 10 million Roma family-oriented society in Europe, the number of basic society members is only 670,000.

For critical social space, its calculation formula is:

$$V_C = 3Mnv_p = 3Nv_p = 3 \times 10^7 \times 0.06 = 1.8 \times 10^6 \left(m^3 \right)$$

Here v_p is the social space owned by a single social individual, and N is the total number of social individuals in this society.

The formula for calculating critical pressure is:

$$p_C = \frac{a_{Hs} M^2}{27b^2} = \frac{a_{Hs}}{27n^2 v_p{}^2} = \frac{1.5 \times 10^{10}}{27 \times 15^2 \times 0.06^2} = 6.86 \times 10^8 \left(Pa \right)$$

The critical pressure is also relatively high. It shows that Roma can enter a condensed state (liquid state) even under high social pressure conditions. Considering the inherent prejudice against the Roma in Europe, it can be considered that the Roma are indeed under relatively high external pressure in Europe. However, the Roma have a long history of being able to maintain a condensed state of the whole family society at all times by cleverly adopting a mobile approach to relieve external pressures.

From the calculation formula of critical temperature:

$$T_C = \frac{8a_{Hs}}{27\beta_H v_H} = \frac{8a_{Hs}}{27\beta n^2 v_p} = 1.19 \times 10^4 \left(K \right)$$

It can be seen that the critical temperature is still relatively high.

Now let's calculate the social temperature of Roma society. Judging from various evidences, Roma society is most likely to be in a liquid state. Let us first roughly estimate the social space where Roma gather according to the parameters of a social society in which all social individuals are basic society members. Let's first estimate from the perspective of energy consumption. According to the calculation that each Roma consumes 1,500 kcal of energy per day, the total energy consumed by the Roma society is

$$dE_{nutrition} = 1500 \times 4184 \times 10^7 \approx 6.28 \times 10^{13} (J)$$

Of course, since modern society and the migration process of Roma people need to consume more extra energy, the extra energy that needs to be consumed will account for a slightly higher proportion of nutritional energy. Take 60% here. Then the social temperature of Roma society is

$$dT = \frac{dU}{C_v} = \frac{dU}{N\beta} = \frac{1.6 \times 6.28 \times 10^{13}}{10^7 \times 2.76 \times 10^4} = 364(K)$$

On the other hand, because the external pressure can be relieved by means of mobility, the total external pressure on the Roma society as a whole is actually not large. In this way, the pressure of the whole society is mainly caused by internal pressure. We can calculate the social space occupied by the Roma through the social temperature of the Roma as

$$V = \frac{a_s N}{\beta T} = \frac{1.5 \times 10^9 \times 10^7}{2.76 \times 10^4 \times 364} \approx 1.53 \times 10^9 \left(m^3\right)$$

This is roughly equivalent to 510 square kilometers. About the size of a large city. Therefore, the calculation results are quite reasonable.

Then we consider the family aggregation effect of Roma. According to the scale of 15 people in each family, the actual social temperature of the Roma family society can be calculated as:

$$T_{HRoma} \approx \frac{a_{Hs}M^2}{N\beta V} = \frac{1.5\times10^{10}\times6.67^2\times10^{10}}{10^7\times2.76\times10^4\times1.53\times10^9} \approx 16(K)$$

It can be seen that the actual social temperature of Roma society has become very low, only 16K, because of the aggregation effect of families. This is very helpful for the Roma to obtain energy from the environment.

That is to say, despite the larger number of Roma households, which can lead to a significant reduction in the social temperature of Roma societies, they are able to maintain a high critical temperature. This also allows the Roma society to maintain the form of a liquid society even when the environmental resources are not very sufficient.

Of course, from the perspective of Roma social temperature, if the external social pressure of Roma is very small, according to formula (8-9), although the critical temperature of Roma is lower than that of social-oriented society, their social temperature will be reduced by the same percentage. Therefore, the entire Roma society can still normally live in its own way.

However, if the external social pressure becomes very large, according to formula (8-12), it can be seen that the social temperature of the Roma will be raised to almost the same as the social temperature of other surrounding ethnic groups.

Which is

$$T_{HRoma} = \frac{a_s N}{\beta V} = 364(K)$$

This temperature is still far from the critical social temperature of 11,900K. However, relative to the higher critical temperature of the surrounding mainstream society, Roma society will definitely become less stable. Of course this change does not seem to be enough to completely break down the condensed state of Roma society.

This was evident in Eastern Europe during the Soviet era. Due to the coercive policies in Eastern Europe at the time, Roma were unable to migrate and had to work with mainstream ethnic groups and their children had to go to school (Marushiakova-Popova & Popov, 2019). This makes the Roma society face unprecedented social pressure, resulting in obvious changes in the traditional social form of the Roma. On the surface, the Roma have successfully integrated into mainstream society. However, in follow-up studies, some studies have suggested that this drastic social change has caused alienation in the relationship between Roma and non-Roma (Bartash, 2015).

Judging from the actual changes in Eastern Europe, this is indeed the case. In Soviet times, on the surface, the Roma were indeed integrated into the entire Soviet society. There is a similar situation in Czechoslovakia in Eastern Europe. However, after 1989, Czech policy towards Roma has changed a lot. Restrictions on various policy aspects for Roma have been eased. Judging from the current situation, it seems that the Roma in the Czech Republic have returned to the original state of no settlement in the traditional sense, and the relationship between the Roma and the mainstream ethnic groups has indeed become more alienated.

That is to say, although the high-pressure ethnic policy of the Soviet Union can temporarily suppress the Roma into the mainstream society, the family atom structure of the Roma has not been destroyed. Therefore, after the pressure on this policy disappears, the Roma's family atom can also continue to form a covalent bond with other family atoms and restore it to the Roma culture with a tradition of more than a thousand years.

Given the very stable social status of the Roma at present, various social problems may arise. In particular, conflicts with mainstream ethnic groups have led to conflicts between some ethnic groups. At present, the best situation in Europe should be to integrate the Roma into European society as much as possible.

There are many ways to achieve this integration of Roma into mainstream European society. Of course, adopting the high-pressure policy of the Soviet Union can indeed raise the social temperature of the Roma, so on the surface it seems that the Roma are integrated into the mainstream society. But at least now it seems that the various side effects caused by this high-handed policy are very large. For example, after this unsustainable high-pressure policy disappeared, the relationship between the Roma and the mainstream ethnic groups became more distant.

In fact, in order to better integrate the Roma into mainstream society, there are still many methods that can be adopted. The main purpose of these methods is to raise the social temperature of the Roma society, so that the Roma society can enter a gaseous society. This makes it easy to integrate into mainstream society.

According to formula (8-5), it can be seen that reducing the social interaction coupling coefficient may one of the effective ways.

For example, if the coefficient of social coupling between families in Roma society is reduced by a factor of 100, then

$$T_C = \frac{8a_{Hs}}{27\beta n^2 v_p} \times 0.01 = 119(K)$$

The critical temperature becomes very low.

Considering the strong interaction coupling coefficient within Roma society, it is mainly because Roma society is a very traditional family-oriented society. Therefore, maintaining a high-strength interaction coupling coefficient mainly relies on gender interaction. That is, intermarriage between families forms a covalent bond. This is of course a traditional cultural practice that deserves respect. But we can also notice that social interaction in family-oriented society also relies on the "ionic bond" between family atoms. That is, through social division of labor, there is still a lot of life in a family that needs to be completed by professionals after all.

In traditional society, only some craftsmen can realize this social division of labor. The Roma themselves are a people who can sing and dance, so they are more suitable to take on the role of this social "ionic bond" in the traditional society. On the contrary, craftsmen in the mainstream society cannot enter the Roma society to perform services. However, the current industrialized social changes have led to a more and more detailed division of labor in society. This also means that even for Roma to live a traditional life, more professionals are needed to complete the service. Therefore, by increasing the opportunities for Roma to receive social services and serve the society, the interaction between Roma and mainstream society can be effectively enhanced. While the interaction between Roma and mainstream society has increased, the interaction within their society has also weakened relatively.

Second, the application of information technology can increase the social space of individuals, thereby reducing the critical social temperature. For example, by popularizing the use of Web2.0, each person's social space has increased by 100 times, that is, from 0.06 cubic meters to 6 cubic meters. Then the critical social temperature will also drop rapidly to 119K. And unlike reducing the social interaction coupling coefficient, which reduces the social temperature proportionally, it can be found from the calculations in the previous chapters that popularizing the use of information technology can also rapidly increase the social temperature. This makes it easier to quickly turn the whole society into a state of gas.

That is to say, the mainstream society should fund the Roma society as much as possible in the application of information technology. The more advanced the information technology is, the larger the social space of the individual society is, and the faster the critical temperature of the whole society will decrease. The reduction of the critical social temperature is an important prerequisite for the society to enter the gaseous state at a relatively low temperature.

Of course, it can also be seen from formula (8-5), although the increase of the population of a household helps to reduce the critical temperature. But it will also lower the social temperature. This doesn't do much for a family-oriented society to get out of a condensed state. On the contrary, the reduction in the number of family members helps to improve the educational level and further increase the information temperature of the whole society. It is more beneficial to the overall improvement of social temperature.

8.3.4 Solid family society

The characteristic of solid-state family-oriented society is that family mobility is relatively weak. Basically, it survives and develops in one area for a long time. This stems from those traditional cultural areas with a long history. However, solid-state family-oriented society is different from solids such as metal, and there is still fluidity within the family.

A solid-state family-oriented society looks more like an organism. The shape can be changed by external force, but it has strong toughness. This is different from a liquid family-oriented society. Solid-state family-oriented society family atoms have very fixed positions and connections with other families. But in liquid family-oriented society, the position between family atoms can easily change. The temperature of solid family-oriented society is lower, and the temperature of liquid family-oriented society is higher.

To form a solid family, of course, means that a very strong social interaction is required. In some towns with a very long history, with a population of several thousand households throughout the town, you will find that all of them seem to be related to each other. This shows that in this small town, all the populations are actually connected to each other. This very strong social connection makes the whole town as a whole, and its solid state is very strong, and to break this connection requires a very strong external force.

In some traditional mountainous areas of East Asia, you can notice that such solid-state towns do exist. Because the difference between liquid family-oriented society and solid family-oriented society is mainly in terms of social temperature and mobility. Therefore, when subjected to external pressure, the liquid family-oriented society can always maintain the integrity of its own family-oriented society, and will not show a reaction force to the external pressure. When a solid-state family-oriented society is subjected to a large external pressure, it will have a large reaction force to the external pressure. And if the external pressure is very large, the solid-state family-oriented society will also adjust its social structure appropriately to adapt to the changes of the new external environment.

References

Ardizzoni, S. (2021). The paradigm of Hakka women in history. *Asian Studies*, 9(1), 31–64.

Bartash, V. (2015). The sedentarisation of Roma in the Soviet Union after 1956: A case study from the former Belarussian Soviet Socialist Republic1. *Romani Studies*, 25(1), 23–52.

Cahn, C., & Guild, E. (2010). *Recent migration of Roma in Europe*. Council of Europe.

Erbaugh, M. S. (1992). The secret history of the Hakkas: The Chinese revolution as a Hakka enterprise. *The China Quarterly*, 132, 937–968.

Guoping, S. (2013). Recent research on the Hemudu culture and the Tianluoshan site. In *A companion to Chinese archaeology*, Blackwell Publishing. 555–573.

Jones, J. (2010). Global Hakka: a case study. *Asian Ethnicity*, 11(3), 343–369.

Kitaoji, H. (1971). The structure of the Japanese family. *American Anthropologist*, 73(5), 1036–1057.

Kumagai, F., & Keyser, D. J. (1996). *Unmasking Japan today: The impact of traditional values on modern Japanese society*. Greenwood Publishing Group.

Kundra, N. (2013). Walking down the lane of history: A study of Nationalism in Bankim Chandra Chatterjee's Anandamath and Rabindranath Tagore's Gora. Creative Zone: Critical Observations on English Literature, 19–21.

Leong, S. T., Wright, T., & Skinner, G. W. (1997). *Migration and ethnicity in Chinese history: Hakkas, Pengmin, and their neighbors*. Stanford University Press.

Li, R. (2021). Focus on important archaeological discoveries and research progress in the Neolithic Age, The State Administration of Cultural Heritage held a working meeting on the important progress of the "Archaeological China" major project. http://www.ncha.gov.cn/art/2021/12/1/art_722_172132.html

London's Population. (2021). London Datastore. https://data.london.gov.uk/dataset/londons-population.

Marushiakova-Popova, E. A., & Popov, V. (2019). Between two epochs: Gypsy/Roma movement in the Soviet Union and in the post-Soviet space. Between the Worlds.

McGarry, A. (2014). Roma as a political identity: Exploring representations of Roma in Europe. *Ethnicities*, 14(6), 756–774.

Mendizabal, I., Valente, C., Gusmão, A., Alves, C., Gomes, V., Goios, A., ... & Prata, M. J. (2011). Reconstructing the Indian origin and dispersal of the European Roma: a maternal genetic perspective. *PloS one*, 6(1), e15988.

Moon, O. (2002). Is the ie disappearing in rural Japan?: the impact of tourism on a traditional Japanese village. In *Interpreting Japanese Society* (pp. 137–150). Routledge.

Seidl, I., & Tisdell, C. A. (1999). Carrying capacity reconsidered: From Malthus population theory to cultural carrying capacity. *Ecological Economics*, 31(3), 395–408.

Sinclair, A. T. (1907). The oriental gypsies. *Journal of the Gypsy Lore Society*; Edinburgh, 1, 197.

Tokyo Population. (2021). World population review. https://worldpopulationreview.com/world-cities/ onfu-population.

Wang. Y, J. (2020). The Enlightenment of Singapore's Family Policy to China (in Chinese). *Youth and Society* (19), 158–159.

9

Balanced society

9.1 The origin and development of balanced society

9.1.1 The family-oriented society of ancient Greece and Rome

From various evidences, ancient Greece was a family-oriented society (Patterson, 2009). The family played a central role in the entire ancient Greek society (Lacey, 1968). This is mainly related to the subtropical maritime climate in which Greece is located. The warm and humid weather all year round allowed the ancient Greeks to give the couple more private space in terms of house structure, so as to meet the requirements of gender interaction of sufficient strength. The sufficient supply of energy in the natural environment made the ancient Greeks not need to migrate frequently to obtain sufficient food sources like the nomads in the north. This guarantees the stability of the family. Like the family-oriented society in Asia, the family in ancient Greece could pass it on to their children because the acquisition of wealth even exceeded their own needs.

The latitude and development history of ancient Rome are basically the same as those of ancient Greece, and it inherited a large amount of ancient Greek culture around 200 BC. Therefore, ancient Rome is also a typical family-oriented society, which contains some important characteristics of family-oriented society (Rawson, 1987).

Of course, both ancient Greece and ancient Rome are closely related to a longer history of ancient Egyptian civilization. Since the rich land of the Nile River Basin can fully meet the energy needs of human existence, both the privatization of property and the privacy of gender interactions are fully guaranteed. This is the basis for the protection of the family-oriented society in ancient Egypt.

Therefore, from the age of European civilization to the demise of Western Rome, the European civilized society was basically a family-oriented society.

Beginning in AD 400, the northern nomadic peoples such as the Germans and Hun people began to invade Western Rome, which eventually led to the demise of the Western Roman Empire.

Now it seems that the strength of the Western Roman Empire is far inferior to that of ancient China in East Asia. After these nomadic people invaded, they basically used their own social culture to control the original Western Roman Empire. Therefore, the kingdoms established at the beginning of the conquest of the Western Roman Empire also belonged to the social-oriented society. But after all, the culture of these nomads is relatively backward, and the Eastern Roman Empire still exists, preserving the essence of the family-oriented society culture of the ancient Roman Empire to the greatest extent.

9.1.2 The emergence of balanced society

In the late Eastern Roman Empire, the Mongols rose and entered Europe, bringing with them the civilization of a social-oriented society from Asia (Morgan, 1986; Jackson, 2014). It has laid a solid cultural foundation for the further development of the social-oriented society in northern Europe. The demise of the Eastern Roman Empire brought the family-oriented society culture of ancient Rome back to

Europe. Combined with the social-oriented society cultural junction that was occupied by the barbarians and was not profound, through the Renaissance, the cultural essence of the family-oriented society was re-inherited. Eventually formed the balanced society culture in Europe.

In the European balanced society culture, countries near the north are dominated by social-oriented society culture and absorb some of the essence of family culture. In the southern countries, the family culture is the main body, and part of the social-oriented society culture is absorbed. Among them, Italy was the core seat of the Roman Empire and the origin of the Renaissance. The most complete family-oriented society culture in Italy is an example.

9.1.3 Diverse society

After entering the modern history of the world in the 16th century, the American continent was discovered by Columbus, and Europe began the era of great navigation. At this time, with the power of the Industrial Revolution, some European countries began to expand their territories. Various colonies were established overseas. Since human beings have been fully developed, the earth has basically been inhabited. Wherever there is a population, there is a culture. Of course, there are advanced and backward distinctions.

On the one hand, European colonial rule expanded its territory and provided more energy for industrialization. At the same time, they also spread their own culture to other countries and nations. For example, in the United States, the British balanced society culture has been spread to the past, and at the same time, the traditional culture of the American Indians has been combined. And a large number of black slaves were transported to the Americas, which brought African traditional culture to the Americas. In this way, the United States today is not only a balanced society, but also a pluralistic and multiracial society. It is believed that this will also be an important development trend of cultural integration in the whole world.

9.1.4 Social changes in the Internet era

The emergence of Internet technology has had a significant impact on social change. From our previous analysis of social thermodynamics, the emergence and use of the Internet are mainly parameters that can affect social thermodynamics in two aspects.

One aspect is to expand personal social space. Due to the use of advanced Internet technology, especially web2.0 technology, personal social space has been greatly increased. The increase of an individual's social space can directly affect parameters such as critical volume, critical temperature and critical pressure. These important parameters are the key parameters that determine whether the entire social thermodynamic system can enter the condensed state. It can be seen from the previous calculation examples that the increase of the social space of a single social individual will lead to the increase of the critical volume, the increase of the critical pressure and the decrease of the critical temperature. Among them, the decrease of critical temperature has the most direct effect. When the critical temperature is lowered, it means that the social temperature must be very low so that the entire social thermodynamic system can enter the condensed state. Whether it can enter a condensed state is the most important sign of social stability. If the society remains gaseous, it means that the social thermodynamic system will be unstable. A less stable social thermodynamic system also means that the system can expand or contract easily. If the thermodynamic system expands, it means that there will be a threat to the adjacent society, because it will occupy the social space of the adjacent society. And if the social thermodynamic system shrinks, it means that the pressure inside the social thermodynamic system will increase. At the same time, a gaseous society also means that it is easy to penetrate into other societies, be absorbed by other societies, and eventually disappear.

The other aspect is to changes in social interaction between social individuals. If it is in a social or balanced society, the changes in social interaction caused by information technology can either enhance social interaction or weaken social interaction. It mainly depends on how these information technologies are used. However, in the family-oriented society considered, the social interaction is mainly caused by gender interaction. The social interaction caused by information does not account for a large proportion in the family-oriented society. Therefore, whether information technology weakens social interaction or enhances social interaction, the impact on family-oriented society is not as great as that of social-oriented society or balanced society. In comparison, the control of the use of information technology, or the control of information, will have a greater impact on social-oriented society than on balanced society.

But the considered information technology really has a very large impact on the critical temperature, which makes even the family society destroy the stability of the family society due to the decrease of the critical temperature. Therefore, the impact of information technology on the stability of family-oriented society is still very large. If the use of information technology causes the family form to become a gaseous state, it means that social individuals will continue to be evaporated from the family society and become separate and independent social individuals. In this way, there will be more social individuals in the whole family-oriented society as members of the basic society. Therefore, with the widespread application of information technology, especially Web2.0, family-oriented society tends to change to a balanced society. This is also the reason why there are more and more balanced societies in the world.

At the same time, some traditional family-oriented societies are still relatively backward in terms of production methods, so due to economic pressure, individual members of some traditional family-oriented societies will flow into the developed balanced society economy. Then, through the information interaction between individual members of the society, various information of the balanced society is transmitted back to the family-oriented society, which will also promote the transition from the family-oriented society to the balanced society to a certain extent.

A similar situation arises for social-oriented society. Therefore, in general, social-oriented society and family-oriented society all over the world tend to transition to balanced society.

9.2 Interaction of balanced society

9.2.1 The interaction between social individuals and family atoms

In a balanced society, there are almost equal numbers of gender molecules and family atoms, which means that the social interaction between gender molecules and family atoms needs to be considered.

For social individuals, the social interaction between them is mainly realized by the way of information exchange. Family atoms mainly generate social interaction through marriage and other forms related to gender interaction.

In the traditional family-oriented society, it is very difficult for such an independent social individual to directly interact with the family atom by means of information transmission. However, in an industrialized society, due to the differences in social division of labor, to maintain the integrity and stability of family life, relevant professionals are required to participate. This significantly increases the chances of social individuals interacting with family atoms in a family-oriented society.

Therefore, in a balanced society, there are also opportunities for social individuals to interact directly with a family atom due to the difference in social division of labor.

The interaction between the social individual and the family atom is actually the same as the interaction of the ionic bond type in the family-oriented society. It can also enhance the interaction between family and family to a certain extent.

But we can also notice the social interaction caused by this ionic bond in family-oriented society, its purpose is to enhance the interaction between families and families. In a balanced society, these professionals are also independent social individuals, and because of the large number of these social individuals, there is also a very strong social interaction between them. This means that the interaction between social individuals and the interaction between families are equal.

In a balanced society, this kind of social individual serves the family atom, which is essentially the social interaction or information interaction between the social individual and the family atom. This is the same in nature as the professional services provided by professionals to a basic society member in a social-oriented society.

9.2.2 Interaction between gender molecular and family atom

In a social-oriented society, a single gender molecule looks a lot like a family. But in fact it is very different from the family atom in the family society. There are mainly the following two aspects:

First, in a social-oriented society, some gender molecules do not have children, or are only a temporary combination. Not as solid as the family atom. Therefore, in a balanced society, the interaction between the gender molecule and the family atom is completely consistent with the interaction between a single social individual and the family atom.

The second is some gender molecules. Although there are children and even relatives and friends, the kinship between such children or relatives and friends does exist and is even very strong. But this kind of relationship, in social-oriented society, only exists as a relatively strong social interaction. They are an integral part of the interaction of the whole society. The children or kinship in the family atom belong to the interaction within the family. In a family oriented society, the social interaction within the family is far stronger than the social interaction among children or relatives in a social-oriented society. Therefore, in a balanced society, it is difficult to tightly combine gender molecular and family atom through this covalent bond. Even in combination, two situations may arise. One is that the young men and women have become a gender molecule independently and become a member of the social-oriented society. The second is that gender molecular accepts the culture of family atoms and is willing to form a closer social interaction within the family, so that a covalent bond between the two atoms can be formed.

It can be seen that in a balanced society, the interaction between basic society members has become more complicated. Some of these social interactions will be very strong while others will become looser. This also leads to certain difficulties in the complete integration of different cultures, or it takes a longer time to promote the integration of two different social cultures.

The diverse cultures that currently exist in European and American countries can be used as strong evidence for the difficulty of cultural integration. Of course, this also includes the integration between the Roma and the mainstream European ethnic groups. The entire fusion process has not even been completed for more than 1,000 years.

9.3 The nature of interfaces between multiple societies

9.3.1 Multiple types of social contacts

There is a unique phenomenon in balanced society, that is, two types of social groups will be closely combined in physical space. But considering the analysis we did earlier, including interactions between social individuals and family atoms, and gender molecules and family atoms, there are significant barriers to these interactions. So it is very difficult for these two types of society to merge together at once.

Since the two social types cannot be completely integrated, both social types will have their own social circles, and in the space of their own social type, they will live completely according to their own way of life. But there will still be various interactions between the two societies, the most important of which is the mutual exchange of information and matter resources. It even includes interactions between people. Gender interactions between social individuals may also be included. To realize these exchanges and interactions, the interface between the two societies must be penetrated.

This is very similar to a semiconductor *PN* junction. In Chapter 2 we have analyzed the formation of such interfaces. This relationship between thermodynamic parameters near the interface involves a non-equilibrium thermodynamic problem. On the one hand, due to the existence of a certain culture on one side of the interface, the number of social individual members of this culture is very large and the concentration is very high. At this time, there is a tendency to spread to another society with low concentration. Another culture with a high concentration on the other side will also spread to this side. This creates a special area on both sides of the interface. Two or more social types of cultures coexist in the region. Individual members of society can live together peacefully without cultural conflict. This is an area where a multicultural society coexists. Such is the case with Kashmir in Asia, for example. Although three of the different cultures belong to the family-oriented society. But their religious beliefs are very different, namely Hinduism, Buddhism and Islam. However, in the course of nearly a thousand years of historical development, this region has always maintained a very stable social environment. Various cultures interpenetrate each other, and they are different. There is no question of changing the other's culture.

9.3.2 Space energy region

Figures 2-6 form two social interfaces. After a sufficiently long diffusion of social individuals, a special area is formed on both sides of the contact surface, which is called "cultural junction" in Section 2.4.3.

As with the *PN* junction, multiple social types or cultures coexist in the areas on either side of the cultural junction. The phenomenon of the simultaneous coexistence of this multi-society or multi-culture is caused by the "cultural gradient" formed by the "high concentration" of specific cultural individuals near the cultural junction.

Since it is very similar to the *PN* junction, we can also use some methods in semiconductor physics to deal with this cultural junction problem.

Now suppose two different cultures come into contact. The two cultures are A culture and B culture. The social individual densities of the two cultures are $N_A$and N_B, respectively, and the unit is m^{-3}.

Of course, since the two cultures are closely intertwined, there must be a small number of individuals of the other in one dominant culture. For the convenience of handling the problem, we can call the social individuals of the mainstream culture in the society as "majority individuals", and the social individuals of non-mainstream culture in the society as "minority individuals".

The relationship between cultural junctions, majority individuals, and minority individuals is shown in Figure 9-1. Among them, the A-culture social individuals are represented by white circles. The B-culture social individuals are represented by black circles.

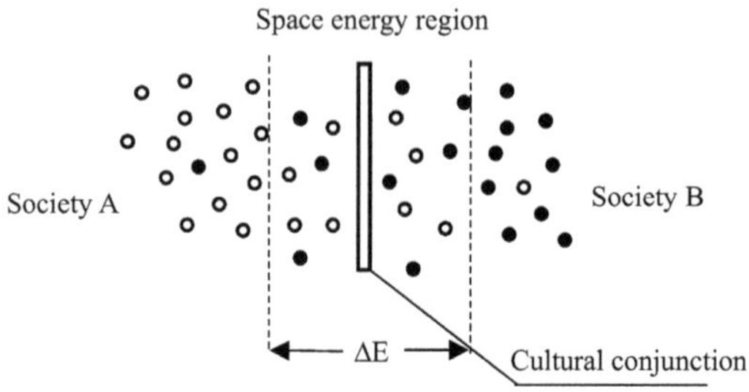

Figure 9.1. Culture conjunction and space energy region

According to the conclusion of semiconductor physics, in the society-A area of cultural junction, majority individuals are A individuals, while minority individuals are B individuals. Conversely, in the society B area, majority individuals are B individuals, while minority individuals are A individuals.

As can be seen from Figure 9-1, since the density of A individuals in society A is very large, while there are almost no A individuals in society B, there is a tendency for A individuals to spread to society B. Conversely, individual B tends to spread to society A.

When individual A spreads to society B, the number of individuals A in society B on the other side of the interface gradually increases, and the density of individual A begins to increase. The diffusion of individual B to society A is similar.

When individual A spreads to society B, it will lead to the coexistence of the two cultures near the cultural junction in society B, and the social characteristics of the cultural junction area will weaken.

We assume here that a society has a need for integrity. This integrity implies the ability to self-repair damaged structures if there is a situation where they may be damaged. Therefore, when more A individuals spread to the B social area, the B social type in the area is destroyed. So there will be an energy of social exclusion, preventing A from entering Society B. The more A individuals enter the Society B, the greater this repulsive energy will be, and it will prevent more A individuals from entering the Society B.

The same is true for the diffusion of B individuals into Society A. There will also be energy that society A rejects individual B from entering. The greater the number of B individuals entering the Society A, the greater this social exclusion force will be.

When a stable state is finally reached, a "space energy region" is formed at the cultural junction. The existence of this space energy region will exclude the social individuals of the other party from entering their own society. However, it can attract individuals belonging to this society to return to the society.

Math further

Assuming that on the side of society A, according to semiconductor physics, the densities of the two social individuals in the space energy region conform to the Boltzmann distribution, so the following two formulas can be used to calculate.

The density of social individual A is:

$$N_A = N_0 e^{\frac{E_A}{k_B T}} \tag{9-1}$$

The density of social individual B is:

$$N_B = N_0 e^{-\frac{E_B}{k_B T}} \tag{9-2}$$

In the formula, N_0 is the initial density of the corresponding social individuals before the area has no contact. It is assumed that the initial density of the two societies is the same. E_A and E_B are the energy of the space energy region, respectively. The higher the energy of the space energy region, the stronger the ability to exclude the other social individuals from entering. Also k_B is the Boltzmann constant and T is the temperature.

The sign of space energy reflects the relative situation of the energies possessed by two societies. That is to say, if the energy of one society is defined as positive energy, the other energy can be defined as negative energy. The definition of this positive and negative energy sign is arbitrary. There are no other sociocultural implications.

In this way, we can find the energy difference on both sides of the space energy region as:

$$\Delta E = E_A - E_B = k_B T ln \frac{N_A N_B}{N_0^2} \tag{9-3}$$

This shows that if the number of two kinds of social individuals near the cultural junction increases, it will lead to an increase in the energy difference of the spatial energy area. Thus, it will be easier to prevent the other social individuals from entering their own society. Similarly, the higher the social temperature *T*, the greater the energy difference will be, and the greater the resistance to the social individuals of the other party from entering their own society.

Of course, if the external energy can also be input. If the energy sign of the external input is the same as the energy difference shown in formula (9-3), the energy of the space energy region can be enhanced. This will be able to more effectively prevent the entry of the other social individuals. On the contrary, it will make it easier for the other social individuals to enter their own society. This external energy is mainly related to social policies and laws. It is also related to the mutual friendliness between the two societies. If the two societies are hostile, the energy difference in the spatial energy zone can be enhanced, thereby preventing social individuals from entering each other's society. And if two societies are friendly, social individuals are more likely to enter each other's society.

9.3.3 Case study: A quantitative analysis of the academic differences of students from Chinese and Western cultural Backgrounds

In Section 3.5.3, we qualitatively analyze the influence of Chinese and Western cultures on the academic achievement of adolescents. Now that we have quantified social thermodynamic parameters, we can quantify some of these differences. In this way, a more accurate relationship among the thermodynamic parameters can be obtained.

Now we assume that such a school exists in a balanced society with a campus area of 0.16 square kilometers. Among them are 1,000 mainstream Western students and a sufficient number of Chinese students. The "sufficient number" here means that they can form a relatively independent thermodynamic system with statistical effects.

For Chinese students, the number is $N_c = 20$. It is calculated according to the nutrition that each social individual obtains 1500 kcal calories.

$$dE_{nutrition} = 1500 \times 4184 \times 20 \approx 1.26 \times 10^{8}\left(J\right)$$

Teenagers need to absorb more high-calorie nutrients such as protein and fat. At the same time, modern food can also carry out deeper processing of food. These all require extra energy. Therefore, the proportion of extra energy consumption of adolescent students will be higher than that of adults. Here it can be calculated according to the proportion of 70%, so that the social temperature of these Chinese students can be calculated as:

$$dT = \frac{dU}{N_c\beta} = \frac{1.7 \times 1.26 \times 10^{8}}{20 \times 2.76 \times 10^{4}} = 387\left(K\right)$$

According to the equation of state of an ideal gas, we can calculate the external pressure on Chinese students:

$$p_{chinese} = \frac{N_c\beta T}{V} = \frac{20 \times 2.76 \times 10^{4} \times 387}{0.16 \times 10^{6} \times 3} = 445\left(Pa\right)$$

Although the social interaction among Chinese students is not great. However, because it is in a balanced society, traditional Chinese culture does not belong to the mainstream culture in a balanced society. Therefore, in the communication of social individuals, the mutual exchanges between Chinese students will be strengthened accordingly. Therefore, it can still be calculated according to the normal social interaction coupling coefficient. That is, the interaction coupling coefficient between Chinese students is:

$$a_{cs} = 1.5 \times 10^{9}\, Pa \cdot m^{6}$$

It is assumed here that there is no kinship among Chinese students. Because for a family-based society, kinship will lead to a rapid increase in the interaction coupling coefficient between them.

In this way, the internal pressure of Chinese students is

$$p_{ichinese} = \frac{a_{cs}N_c^2}{V^2} = \frac{1.5\times10^9\times20^2}{\left(0.16\times10^6\times3\right)^2} = 2.7\left(Pa\right)$$

It can be seen that in this school, the internal pressure is far less than the external pressure. So it is more likely that Chinese students are actually in a gaseous state. The gaseous state means that this group of Chinese students is easier to integrate into the mainstream culture of the entire campus.

Of course, it can be seen from the calculation that if the interaction coupling coefficient between Chinese students is enhanced, the internal pressure can increase rapidly. The reason for this enhanced interaction may be that there are specialized organizations to enhance mutual connection among students, etc.

For 1,000 Western students, the composition may also be more complex. However, all of them are considered in terms of social individuals in a social society. If nutritional intake were the same, the social temperature of these students would be the same.

However, due to the increase in the number of people, the external pressure and internal pressure will be different.

We can calculate the external pressure on all other students:

$$p = \frac{N\beta T}{V} = \frac{1000\times2.76\times10^4\times387}{0.16\times10^6\times3} = 2.22\times10^4\left(Pa\right)$$

$$p_i = \frac{a_sN^2}{V^2} = \frac{1.5\times10^9\times1000^2}{\left(0.16\times10^6\times3\right)^2} = 6680\left(Pa\right)$$

It can be seen that the difference between external pressure and internal pressure of all Western students is not as large as that of Chinese students. Therefore, due to internal pressure, Western students in this school are more likely to form a condensed state.

Of course, if all students, including Chinese students, can fully communicate with each other, the gaseous Chinese student group can easily integrate into the condensed mainstream student group.

But we can also notice the cultural background differences.

One of them is that in terms of social space, the social space of individual Chinese students is smaller than that of Western students. So if Western students $v_p = 1m^3$, then we can set Chinese students $v_{pc} = 0.5m^3$

In addition to the same external pressure as Western students on campus, Chinese students also experience strong internal pressure from their families p_{home}. There is still a lot of pressure within the family. In some families in China that attach great importance to their children's education, they can let their children get very good scores in the college entrance examination in a school with a poor reputation. And turn a school with a poor reputation into a school with a very good reputation. Therefore, the pressure within the family can be set to be twice that of the school. That is, $p_{home} = 2p_{chinese}$. These pressures combine with the pressure on the campus to form a resultant force.

So the total pressure

$$p = p_{home} + p_{chinese} = 1335(Pa)$$

In addition, considering factors such as cultural background, the social interaction between Chinese students has weakened, but on the other hand, considering that as a minority, under the requirement of social equality, there will be relevant organizations to strengthen the relationship between Chinese students, then take $a_{cs} = 0.5 \times 1.5 \times 10^9 Pa \cdot m^6$

The reduction of social interaction directly leads to the reduction of internal pressure among Chinese students. This effectively means that Chinese students are less likely to enter a condensed state. Therefore, it can be solved directly according to the ideal gas equation of state.

$$T = \frac{(p_{home} + p_{chinese})V}{N_c \beta} = \frac{3p_{chinese}V}{N_c \beta} = 1160(K)$$

At the same time, we can also calculate the critical temperature and critical pressure of Chinese students as

$$T_{Cchinese} = \frac{8a_{cs}}{27\beta v_{pc}} = \frac{8 \times 0.5 \times 1.5 \times 10^9}{27 \times 2.76 \times 10^4 \times 0.5} = 16500(K)$$

Compare the critical temperature of 1,000 other students at this school

$$T_C = \frac{8a_{cs}}{27\beta v_{pc}} = \frac{8 \times 1.5 \times 10^9}{27 \times 2.76 \times 10^4 \times 1} = 16500(K)$$

The critical temperature is the same. This is because with a reduction in the coupling coefficient of social interactions, there is a corresponding reduction in individual social space.

It can be seen from the above calculations that under relatively large family pressure, the social temperature of Chinese student groups is very high. Of course, such a high temperature also means that the independent ability of Chinese students is relatively weak. Active support from the family is required to ensure adequate energy from the surrounding environment. On the other hand, such a high social temperature also means that its internal energy is relatively high. Of course, these internal energies do not fully reflect the neural network energy consumption of the brain. Because the internal energy of social individuals also includes the energy consumption of physiological kinetic energy and the energy of the neural network to deal with other problems. The high social temperature of Chinese students means that most of the energy will be concentrated on the energy consumption of the neural network of the brain required to process a certain task. The task may be to learn all kinds of knowledge. In physical fitness and in the areas of creativity, divergent thinking, mutual cooperation, etc. that require energy consumption, no more energy can be allocated in the past. It is possible that Chinese students will have lower requirements in terms of cooperative exploration, creativity, survival ability, artistic ability, and physical fitness. Fortunately, many parts of East Asia have already noticed this problem. Corresponding educational reform measures have also achieved remarkable results.

Of course, due to the high social temperature, it is naturally more difficult to form a condensed state. This in turn affects the ability of Chinese students to cooperate with each other.

It can also be seen from this case that the differences between Chinese and Western students are quite obvious. The advantages and disadvantages of each are clear. The advantage of Chinese students is that they can concentrate a lot of energy on knowledge learning. This can provide a more solid knowledge base for them to further study natural science knowledge and enter the society in the future. But the disadvantage is also obvious that the cooperation ability is relatively weak, the interest is relatively narrow, and the development of sports, arts and other aspect skills also has a relatively unfavorable impact. At the same time, the ability to survive independently is weaker than that of Western students of the same age.

For Western students, the opposite is true for Chinese students. Western students have a strong ability to survive independently and develop better in the field of sports, arts. In addition, the ability to cooperate is stronger than that of Chinese students. The disadvantage is that the knowledge base is not very solid. This may affect the further study of natural science knowledge in the future, and there will be certain obstacles in acquiring new knowledge in the face of rapid technological progress.

The solution to the problem is actually very simple, that is, to promote closer exchanges between Chinese and Western education. It can not only give full play to the advantages of traditional Chinese culture in imparting knowledge, but also promote the development of Chinese students' cooperation and other necessary abilities.

9.4 Types of balanced society

9.4.1 Gaseous balanced society

In a balanced society, the interaction between social individuals is between that of a social-oriented society and a family-oriented society. Therefore, if the multicultural integration is very good in a balanced society, it means that the social interaction will be more balanced. However, if there are relatively obvious social-oriented society groups and family-oriented society groups in a balanced society, there will be a relatively clear interface between these two different social groups. Then the cultural junction is formed, and the space energy area is formed on both sides of the cultural junction. This is relatively common in some balanced societies.

When the overall social interaction within a balanced society is relatively weak, a gaseous balanced society may appear. Of course, this gaseous balanced society is also not stable. Therefore, if the balanced society is in a gaseous state, it means that a more chaotic situation is likely to occur. For example, during World War I and World War II, there was a very chaotic situation in the balanced society of Europe. This situation is basically the same as the gas state.

But as long as the structures such as family atom, gender molecular, etc. are not destroyed, once the gas state condenses, it will return to the condensed state of the original balanced society.

However, because different types of societies have different critical temperatures, this means that when a gaseous equilibrium society gradually condenses, societies or nations with a relatively high critical temperature will first form a condensed state.

In this way, the balanced society that finally forms a condensed state will be divided into several different types of society.

For example, after the end of World War II, Europe was divided into two camps. Even within a camp there are different social types. For example, Italy, etc. belong to a relatively solid family-oriented society, while Germany, etc. belong to a more typical social-oriented society. The Roma also continue to maintain their own way of life. Although they belong to the socialist camp, there are also substantial differences in social form between Eastern European countries and the republics within the Soviet Union. When the Soviet leaders adopted a series of coercive policies in order to unify everyone's ideology, it led to many contradictions, such as the uprising in Hungary and the Solidarity trade union incident in Poland, etc.

9.4.2 Liquid balanced society

Liquid balanced society is a fluid state. But unlike liquid social-oriented society and liquid family-oriented society, liquid balanced society is a social form in which many different types of fluids coexist. It is like the mixing of oil and water in the physical world. Although the two fluids can briefly mix together to form an emulsion state. However, under the action of external forces, once stabilized, the two fluids can still be clearly separated.

The main reason for this phenomenon is the existence of different types of cultural phenomena in balanced society. The social interaction within different social cultures will be different. Especially the difference between family-oriented society and social-oriented society is very big. The social interaction of family-oriented society is closely related to gender interaction. The social interaction within the social-oriented society is closely related to the transmission of information. This also determines that the atomic and molecular structures of these two different types of societies are very different. Just like the interaction between different types of physical molecules or atoms may be difficult, the interaction between basic society members of family-oriented society and social-oriented society is too different. This makes it difficult for the two type fluids of social groups to mix completely.

9.4.3 Mixed state of balanced society

Of course in a balanced society. Among them, the family-oriented society type of culture accounts for a relatively large proportion, so in this case, there may also be a mixture of liquid and solid, a mixture of gas and liquid, a mixture of gas and solid, or a mixture of gas, liquid and solid.

This is because family-oriented society can achieve self-sufficiency, so in this case, family-oriented society will appear in a solid state. And if there is a gaseous social culture in this society, in this case there is a mixture of gaseous and solid.

Such mixtures of gaseous and solid states generally have greater instability. For example, when the European colonists first came to the Americas, because the climate in the southern and central parts of the Americas was very warm, it is believed that at this time, the central and southern Americas basically belonged to the Indian family-type society. If these Indian tribes did not move on a large scale, it can be proved that these Indian tribes were actually in a state of solid family society. However, after the European colonists came to the Americas, although their homeland was a balanced society, the only people who came to the Americas were those groups with relatively high mobility, so on the whole, these colonists essentially formed a group social-oriented society. And the colonists are very mobile and in a gaseous state.

When these two different forms of society are combined, the gaseous society can easily penetrate into the family-oriented society. In particular, the gaseous society belongs to the social type society, which has stronger expansion. This will inevitably conflict with family-oriented society. Such conflicts may persist over time. After a long period of mutual integration, the current balanced society in various regions of the Americas has finally been formed.

After the society merges with each other, it is easier to form a mixture of gas, liquid and solid state in the end. In this mixed balanced society, it includes not only the fusion of cultures, the fusion of social types, but also the fusion of races. This shows the typical characteristics of this kind of mixed balanced society in various countries in South America.

9.5 Climate change and human social development

9.5.1 Trends in climate change since the birth of the earth

At present, according to the knowledge of the solar system, the origin of the earth came from a huge explosion of the sun. This massive explosion caused the sun to eject some of that material out. Astronomers speculate that some of the material that was first ejected formed the Oort Cloud, which is now the outermost part of the solar system. The Oort cloud formed by this ejection has reached a distance of about 1 light-year from the sun. It can be imagined that the power of the sun's explosion was quite large.

A subsequent explosion formed what is now the Kuiper asteroid belt, and Pluto, once thought to be the ninth largest planet in the solar system, is a dwarf planet in the Kuiper asteroid belt. Of course, what was ejected by the explosion of the sun was gas, and after cooling, a large number of asteroids in the current Kuiper belt were formed. And some of the asteroids, such as Halley's comet, will still orbit the old sun according to a certain cycle.

10 million years after the sun appeared, a big bang from the sun ejected heavier elements that formed the eight planets we know today. Earth is one of them.

It can be speculated that the sun ejected a hot gas when it exploded that year. And this hot gas, after 200 million years of cooling, gradually formed a solid rock state like the earth. After another 800 million years or so, the temperature of the Earth's surface gradually cooled to a level suitable for the emergence and evolution of life.

The earliest ones were very primitive shellfish in the ocean. At this time, the temperature of the earth is very high, reaching an average temperature of tens of degrees Celsius.

However, as the temperature of the earth continued to decrease, about 800 million years ago, the temperature of the earth reached a temperature suitable for the survival of higher organisms. By 300 million years ago, large creatures such as dinosaurs began to appear.

Of course, the evolution and survival of large creatures such as dinosaurs on the earth also caused the temperature of the earth to be more suitable for living creatures to maintain for two to three hundred million years. During this period, the temperature of the earth reaches an average temperature of about 20 degrees Celsius. But as the Earth's temperature continued to cool, the energy of solar radiation caused by asteroid impacts and large-scale volcanic eruptions was blocked, thus causing the Earth to enter a relatively cold glacial period. During such glacial periods, creatures such as dinosaurs could not adapt and eventually became extinct. In their place are mammals that were able to maintain their body temperature during glacial times and hatch offspring inside their bodies. Anthropocene is one of them.

Of course, from the perspective of the evolution of the solar system, the overall temperature of the earth is constantly cooling down. But this cooling time span is in units of hundreds of millions of years. Therefore, for only a few million years of human evolutionary history, the overall temperature of the earth has remained relatively stable.

However, with the movement of the earth's crust and changes in solar radiation energy, the earth also has fluctuations in atmospheric temperature during the Quaternary glacial period. About 10,000 years ago, the Earth ended the 100,000-year Little Ice Age of the fourth glacial period and ushered in a warmer period (Batchelor et al., 2019). During this warmer period, the sea level rose to its current height, and the earth's environment was able to provide a very rich energy for human society, which was established and developed rapidly. The family-based society in human society was first established. But about 2,000 years ago, Earth's temperature dropped slightly. From the perspective of this book, this leads directly to the emergence of a social society. Many animals living in the north have been mass extinct. About 400 years ago, in the last years of the Ming Dynasty in China, the earth ushered in a cooling, which seemed to herald the arrival of a small ice age that may last for a long time in the fourth ice age.

9.5.2 Energy generated by human activities

After thousands of years of development, human society on the entire earth has now become a super-large society with multiple races and multiple social types. This society as a whole can be understood as a balanced society. Let us first calculate some important social thermodynamic parameters corresponding to human society on earth.

Current global population

$$N = 7.2 \times 10^9$$

total land area of the earth

$$s = 1.5 \times 10^8 km^2 = 1.5 \times 10^{14} m^2$$

Total area of Earth's surface including oceans

$$s_{total} = 5.1 \times 10^8 km^2 = 5.1 \times 10^{14} m^2$$

the total mass of the earth's atmosphere is about (Verniani, 1966; Trenberth, 1981, 1994)

$$m_{atm} = 5.14 \times 10^{18} kg$$

Heat capacity per unit air mass

$$C_{atm} = 1006 Jkg^{-1}K^{-1}$$

The daily amount of energy radiated by the sun to the earth

$$dE_{sun} = 1.5 \times 10^{22} J$$

Earth's daily electricity production

$$dE_{electric} = \frac{27 \times 10^{3} \times 3.6 \times 10^{15}}{365} = 2.66 \times 10^{17} J$$

The generation of this electricity includes a large part of coal-fired power generation. Coal is very inefficient, so as much electricity is produced, as much heat is generally produced at the same time. And a lot of electricity is sent to cities and factories, and a large part is also to drive the heat engines, such as the operation of air conditioners. The efficiency of the heat engine is only about 50%. Of course, these powers also include some so-called green energy sources such as hydropower and wind power that do not generate heat. So on the whole, a lot of heat is also generated when this electricity is generated. These calories can be calculated as about 50% of the energy.

At present, the natural energy sources on the earth are mainly the energy of solar radiation and the energy released by the continuous cooling of the earth's interior. These two energies are currently in very short-term equilibrium. But the energy released inside the Earth is slightly larger. This will cause the earth to continue to cool. But after a few billion years, the sun began to cool and became a red dwarf as the fusion fuel was almost exhausted. At this time, the sun will begin to expand, and the energy radiated to the earth will increase rapidly, which will cause the temperature of the earth to rise.

Of course, at present, the energy of solar radiation is still very large. Taking the moon as an example, its interior generates less heat, but its surface temperature can be maintained at an average temperature of about minus 20 degrees Celsius depending on solar radiation. The average temperature of the earth's surface is 15 degrees Celsius.

Of course, the earth does not absorb all the energy of solar radiation. In terms of the proportion of energy consumption, a large part of the solar radiation energy will be reflected back into the universe. This energy is about 34%. That is to say, only 66% of the sun's radiant energy actually hits the earth. Of course, part of the energy radiated to the earth must be absorbed by green plants. These green plants include those on land as well as those in the ocean. According to the global green plant coverage can reach 40%. That is to say, 40% of the energy that the sun really shines on the earth is to be used by green plants for photosynthesis. Due to the high photosynthesis efficiency of green plants, very little heat is generated.

It is possible that the remaining 40% of the solar energy is directly absorbed by the earth's crust, ocean or atmosphere. Due to the high transmittance of the atmosphere, the atmosphere absorbs less energy. In this way solar energy mainly heats the earth's crust and oceans. The heat from the crust and oceans then diffuses partially into the atmosphere, creating a temperature gradient in the atmosphere. Of course, the solar energy that heats the crust and the ocean is more diffused into the interior of the crust and the depths of the ocean. After all, the thermal conductivity of rock is 50 times that of the atmosphere. In this way, we can make a simple estimate, that is, the energy of solar radiation, which is used to heat the crust and oceans, is 50 times the heat that heats the atmosphere. That is to say, only about 2% of the solar radiation energy absorbed by the crust and ocean will eventually be released into the atmosphere, causing the atmospheric temperature to rise. This part of the energy is approximately:

$$dE_{sun2atm} = 0.02 \times 1.5 \times 10^{22} = 3 \times 10^{20} \, (J)$$

Then we calculate the total energy consumed by human activities in detail. In addition to the total global power generation, another energy consumed by human society is mainly natural gas. At present, the amount of natural gas consumed by humans every year is about 4 trillion cubic meters. Calculated according to the combustion value of natural gas of $10000kcal/m^3$, the energy generated by human burning natural gas every day is:

$$dE_{gas} = \frac{40000 \times 10^8}{365} \times 10000 \times 4184 = 4.58 \times 10^{17}\,(J)$$

Except for a small part used for power generation, this part of natural gas is basically used to generate heat.

There is also the burning of oil. Petroleum is mainly used to refine the fuel used in internal combustion engines such as gasoline and diesel. So part is used to generate power and part is used to generate heat. According to the data of BP Statistical Review of World Energy 2017, the world consumes about 96.56 million barrels of oil every year in 2016. Calculated according to Barrel of oil equivalent (BOE), 1 BOE ≈ 6,000 cubic feet of natural gas, that is

$$1BOE = 6000\,ft^3 = 170m^3$$

So the energy consumed by 96.56 million barrels of oil is

$$dE_{oil} = 9.656 \times 10^7 \times 170 \times 10000 \times 4184 = 6.86 \times 10^{17}\,(J)$$

However, considering that the fuel is mainly consumed in the form of a heat engine. According to the heat engine efficiency of 50%, only $3.43 \times 10^{17}\,J$ of energy is actually converted into heat.

This way we can calculate the total amount of heat generated by human activity as:

$$dE_Q = 50\% dE_{electric} + 50\% dE_{oil} + dE_{gas} \approx 9.35 \times 10^{17}\,(J) \qquad (9\text{-}4)$$

It can be noticed that, unlike the solar radiation only a small part of the energy enters the atmosphere, all the heat generated by human activities is all released into the atmosphere.

We can also compare the heat generated by human activity with the heat released into the atmosphere by solar radiation. The proportion of heat produced by human activities is

$$\eta = \frac{dE_Q}{dE_{sun2atm}} \times 100\% = 0.3\% \tag{9-5}$$

It can be seen that this proportion is relatively large. That is to say, the heat generated by human activities is enough to affect the changes in atmospheric temperature.

9.5.3 Effects of heat generated by human society on the atmosphere

We can calculate the additional impact that the heat generated by human activity may have on atmospheric temperature on a daily basis.

$$dT_{atm} = \frac{dE_Q}{C_{atm} m_{atm}} = \frac{9.35 \times 10^{17}}{1006 \times 5.14 \times 10^{18}} = 1.8 \times 10^{-4} (K) \tag{9-6}$$

Then the annual increase in atmospheric temperature caused by human activities will reach:

$$dT_{atmosphere} = 365 \times dT_{atm} = 0.066(K)$$

It can be seen from this calculation result that the temperature rise caused by human activities is actually quite obvious. In a hundred years, it can increase by about 6.6 degrees Celsius. Current data show that the average temperature increase in the Earth's atmosphere over the past 100 years is about 0.8–1.2 degrees Celsius (IPCC, 2022). This is due to the fact that although space is a vacuum, the thermal insulation effect is very good. However, when an object's temperature is relatively high, infrared rays are generated. This infrared light can be radiated out through a vacuum. The energy of this part of the radiation is very large. As a result, the temperature of the Earth's atmosphere will decrease. On the other hand, the energy consumed by human activities in the early stage of industrialization was not as much as it is now, and the energy consumed by human activities has only begun to increase significantly in recent decades. The current study also shows that the temperature change of the earth's atmosphere has accelerated in the last decade (National Academy of Sciences, 2020).

Of course, if it's just the heat from human activity that heats the air, there's a solution. For example, the heat from the power plant is introduced into the earth's crust. However, it can be seen from the above calculations that in fact, the heat released by human activities into the atmosphere is mainly produced by burning natural gas and fuel oil. This part of the energy cannot be imported into the earth's crust. However, even so, the energy released by human activities is limited after all, so eventually the atmospheric temperature is too high, and it will be emitted through the crust and eventually reach thermal equilibrium. After a period of time like this, human development factors will eventually stabilize the temperature of the atmosphere.

9.5.4 Greenhouse gas effects

The heat in the atmosphere can also be radiated into the cosmic space by means of infrared rays, which is a very common way of cosmic space energy radiation. The currently operating Webb Space Telescope uses the method of detecting infrared radiation from galaxies to observe images of distant galaxies. Practice has proved that the effect is much better than visible light.

However, there are also many elements in the atmosphere that have a strong blocking effect on infrared radiation. These elements are called greenhouse gas elements.

Common greenhouse gas elements mainly include carbon dioxide, methane, nitrogen oxides, and water vapor. However, it is carbon dioxide that is currently causing the temperature of the earth's atmosphere to rise. 79% of U.S. greenhouse gas emissions in 2020 are carbon dioxide (EPA, 2022). Other research data show that the concentration of carbon dioxide in the atmosphere is now 40% higher than a hundred years ago (National Academy of Sciences, 2020).

Important evidence to support the greenhouse gas effect now includes the fact that atmospheric carbon dioxide concentrations in the ancient dinosaur era were five degrees Celsius higher than those in the Earth's atmosphere today. The average temperature of the entire Earth's atmosphere in the age of dinosaurs was also much higher than the current average temperature of the Earth's atmosphere. Another evidence is that the atmosphere of Venus is mainly composed of carbon dioxide, and the surface temperature of Venus is as high as 700K or more.

Overall, if this greenhouse gas effect does exist, then it may cause the thermodynamic properties of the entire atmosphere to become very unstable. After all, if the temperature of the atmosphere is increased through the greenhouse effect, it will also increase the emission of greenhouse gases, such as water vapor, and the interaction will cause the temperature of the atmosphere to become higher and higher, and there is no upper limit. The result could be something like the temperature of Venus' atmosphere.

9.5.5 Social Thermodynamic Parameters of Global Human Society

The development of the global society is very unbalanced. The world's most developed G7 countries account for 45% of the world's total economic output and only over 10% of the population. It can be seen that the global economy is mainly concentrated in relatively rich countries. Therefore, in terms of the global average, each individual in society consumes about 1,400 kcal to meet the nutritional needs, the nutritional energy required to maintain the operation of the global society is:

$$dE_{nutrition} = 1400 \times 4184 \times 7.2 \times 10^9 \approx 4.22 \times 10^{16} (J)$$

According to the calculation of other additional energy required to meet nutrition accounting for 50% of the total nutrition, the social temperature of the global human society is

$$dT = \frac{dU}{N\beta} = \frac{1.5 \times 4.22 \times 10^{16}}{7.2 \times 10^9 \times 27600} \approx 318(K)$$

This equates to a temperature of 45 degrees Celsius, which is slightly warmer than the global average. Of course this is the result of global industrialization. At the same time, considering the imbalance of global economic development, the social temperature of some developed countries is much higher, while the social temperature of very backward countries will be much lower. However, such a high average social temperature means that additional energy input through industrialization is required to maintain the normal nutritional needs of the global population. If industrialization stagnates, etc., it may bring about a very serious famine problem. For example, if a large-scale war occurs, industrial production cannot be carried out smoothly, and the global famine problem will appear in a very common form. The blockade of Ukrainian ports by Russia in 2022 has caused a significant increase in food prices in some countries that depend on Ukraine for food imports. If the corresponding measures are not taken in time, the food crisis will really happen.

Next we can also calculate the critical parameters of human society. Considering the unbalanced development of global information technology, here is calculated according to the information space of a single social individual in the world$v_p = 0.1m^3$, then the critical social space of the global society is:

$$V_c = 3\times 7.2\times 10^9 \times 0.1 \approx 3\times 7.2\times 10^8 = 2.16\times 10^9\left(m^3\right)$$

It can be seen that the critical social space is very small, only more than 2,000 square kilometers.

$$p_C = \frac{a_s N^2}{27b^2} = \frac{1.5\times 10^9 \times 7.2^2 \times 10^{18}}{27\times\left(7.2\times 10^8\right)^2} \approx 5.7\times 10^9\left(Pa\right) \tag{9-7}$$

$$T_C = \frac{8a_s}{27\beta v_p} = \frac{8\times 1.5\times 10^9}{27\times 2.76\times 10^4 \times 0.1} = 1.65\times 10^5\left(K\right) \tag{9-8}$$

It can be seen that the critical temperature of global human society is still very high. This also means that from the perspective of the global whole, the global human society can still effectively enter a state of condensed. Of course some of these societies may be in a solid state, and some may remain in a gaseous state.

If the external pressure on the global human society is very small, we can also estimate the size of the social space in which humans actually gather:

$$s=\frac{V}{3}=\frac{a_s N}{3\beta T}=\frac{1.5\times10^9\times7.2\times10^9}{3\times2.76\times10^4\times318}\approx\frac{1.26\times10^{12}}{3}=4.2\times10^{11}\left(m^2\right)$$

This area is about 4.2×10^5square kilometers. Considering that the world's population is mainly concentrated in some large cities, this data is relatively credible. This way we can calculate the internal pressure

$$p_i=\frac{a_s N^2}{V^2}=\frac{1.5\times10^9\times7.2^2\times10^{18}}{1.26^2\times10^{24}}\approx5.02\times10^4\,(Pa)$$

We can also calculate the entropy of human society

$$\begin{aligned}S&=N\beta\left\{\frac{3}{2}lnT+\ln\left(\frac{V}{Nn_p}\right)+\frac{3}{2}\left[\frac{5}{3}+\ln\left(\frac{2\pi mk_B}{h^2}\right)\right]\right\}\\&=7.2\times10^9\times2.76\times10^4\\&\times\left\{\frac{3}{2}ln318+ln\frac{1.26\times10^{12}}{7.2\times10^9\times2\times10^{27}}+\frac{3}{2}\times\left[\frac{5}{3}+ln\left(\frac{2\pi mk_B}{h^2}\right)\right]\right\}\\&\approx3.63\times10^{15}\left(J\bullet K^{-1}\right)\end{aligned}$$

9.5.6 Impact of climate change on human society

1. The impact of external pressure and changes in social space on social temperature

Let us analyze the impact of climate change on thermodynamic parameters related to human society. This assumes that the global average temperature rises by 3 degrees Celsius. After the global average temperature rises by 3 degrees Celsius, the global average temperature will reach 18 degrees Celsius. This is close to global temperatures in the age of dinosaurs [IPCC, 2022], and warmer than global temperatures were more than 2,000 years ago. And that era was a period of rapid development of human civilization.

At this level of global temperature increase, the ice sheets at the North and South Poles will all melt. Sea level will rise by at least 60 m (O'Brien, Robinson, Pancost et al., 2017; Tierney, Poulsen, Montañez et al., 2020). This also means that many low-lying areas around the world will be submerged by seawater.

At the same time, the temperature of the earth has become wetter. Large amounts of water vapor evaporate into the atmosphere, creating a stronger greenhouse effect, which helps keep the planet hot for a long time. Temperatures in the north of the northern hemisphere, such as the Greenland, will become subtropical warm. In this way, energy consumption in this region will be greatly reduced in winter, which will help to slow the rise of global temperature. Of course, the northern hemisphere will gradually shift to a family-based society. It is expected that frequent conflicts within human societies due to the expansion needs of a gaseous society-type society will be drastically reduced. Humanity seems to be on the verge of entering a relatively prolonged period of peace as it was two thousand years ago.

However, the above assumptions may only be possible under very slowly changing conditions. But judging from the current global warming situation, this huge climate change may start to happen in a very short time. In such a short period of time, the entire thermodynamic system of human society has not had time to make corresponding internal adjustments. This in turn leads to the problem that the changes in the global environment are not compatible with the structural adjustment within human society, which in turn causes a very turbulent situation like turbulence in fluids.

First of all, the rise in global temperature leads to the rise of sea level, which brings increased external pressure on the whole society. If human society adds a greater pressure for a short time. This pressure is not negligible relative to the internal pressure. Then the whole van der Waals equation becomes:

$$(p+p_i)(V-b)=N\beta T \tag{9-9}$$

If, on average, each person increases the pressure by $10^{-5}\,Pa$, the total pressure in human society will be:

$$p=7.2\times10^{9}\times10^{-5}=7.2\times10^{4}\,(Pa)$$

Of course, due to the increase of external pressure, it means that more people will actively use information technology to obtain information, so that the social space of individuals will also increase. Assuming that the individual social space increases to $v_p=0.2m^3$at this time, then

$$b=Nv_p=7.2\times10^{9}\times0.2=1.44\times10^{9}\left(m^3\right)$$

If the social space of human society remains unchanged, then we can find that the social temperature of the whole society becomes:

$$T=\frac{(p+p_i)(V-b)}{N\beta}=\frac{\left(7.2\times10^{4}+5.02\times10^{4}\right)\left(1.26\times10^{12}-1.44\times10^{9}\right)}{7.2\times10^{9}\times27600}\approx772(K)$$

It can be seen that the rise in global temperature has brought certain pressures to every social individual. Then this pressure will cause the temperature of the whole society to rise rapidly. The greater the external pressure, for example, the dissemination of information such as earthquakes, tsunamis and other natural disasters and some cities that have already begun to be submerged, will lead to higher social temperatures.

The rise in temperature of the entire human society means that more energy needs to be input to meet the nutritional needs of human beings. This extra energy comes mainly from industrial production. However, if there is chaos in some societies, it may lead to the risk of rupture of some industrial production chains. This further leads to the inability of modern industry to meet the food shortage in local society, and social unrest may occur rapidly.

From the calculation results of formulas (9-7) and (9-8), if the social space of social individuals increases, and if social interactions can also increase synchronously, or even larger, the social critical parameter will be remain unchanged or improve. This contributes to the formation of a condensed state of society as a whole. However, if the social space of individuals increases, but social interactions cannot increase synchronously, for example, consensus cannot be reached on some global issues, the critical parameters of the entire society will decrease. This will make it easier for society as a whole to enter an unstable gaseous state.

But on the other hand, if external pressure increases, the total social space of human beings is reduced. For example, a 90% reduction, then

$$T = \frac{(p + p_i)(90\% \times V - b)}{N\beta} \approx 697(K)$$

It can be seen that the social temperature will decrease. This should reduce the need for additional energy. The main way to reduce total human social space is through greater urbanization. Let the population be more concentrated in some super-large cities, so that the efficiency of human energy utilization will be greatly improved.

This is relatively easy to do when sea levels rise and many coastal areas are submerged. Of course, this requires that the society is sufficiently stable and that inland cities are willing to absorb immigrants from coastal areas.

Also consider:

$$V \gg 10b$$

Therefore, even by using the most advanced network technology to enhance the information space of social individuals, it has little effect on reducing the temperature of society.

2. The influence of social interaction coupling coefficient changes on thermodynamic parameters

Of course, there is another interesting phenomenon in human society, that is, in the face of increasing external pressure brought by various disasters, the interaction between social individuals will be enhanced in a relatively short period of time. It is assumed that under the influence of information that some large cities are submerged, the coupling coefficient of social-individual interaction in human society increases by 100 times in a short time. The improvement of the efficiency of social individuals using information technology has brought the social space of social individuals to $v_p = 1m^3$. Then we can calculate that various critical parameters will become:

$$b = 7.2 \times 10^9 \times 1 = 7.2 \times 10^9$$

$$V_c = 3 \times 7.2 \times 10^9 \times 1 \approx 3 \times 7.2 \times 10^9 = 2.16 \times 10^{10} \left(m^3\right)$$

$$p_C = \frac{a_s N^2}{27 b^2} = \frac{1.5 \times 10^{11} \times 7.2^2 \times 10^{18}}{27 \times \left(7.2 \times 10^9\right)^2} \approx 5.7 \times 10^9 \,(Pa)$$

$$T_C = \frac{8 a_s}{27 \beta v_p} = \frac{8 \times 1.5 \times 10^{11}}{27 \times 2.76 \times 10^4 \times 1} = 1.65 \times 10^6 \,(K)$$

It can be seen that both the critical social space and the critical social temperature have increased significantly. The critical social pressure, on the other hand, did not change. This is because the increase in individual social space offsets the increase in critical social pressure due to the increase in the coupling coefficient of social interactions.

On the whole, this should be more utilized to bring the whole society into a state of condensed. Social individuals are more likely to migrate and concentrate in larger cities.

However, the improvement of social interaction has led to another problem, which is the increase of internal pressure in society. If the social space in which humans gather does not change, then

$$p_i = \frac{a_s N^2}{V^2} = \frac{1.5 \times 10^{11} \times 7.2^2 \times 10^{18}}{1.26^2 \times 10^{24}} \approx 5.02 \times 10^6 \,(Pa)$$

This leads to a rise in social temperature:

$$T \approx \frac{p_i V}{N\beta} = \frac{5.02\times10^6\times1.26\times10^{12}}{7.2\times10^9\times27600} \approx 3.18\times10^4\,(K)$$

This is a very high social temperature. This usually requires higher energy input to maintain. Judging from historical experience, some human societies can maintain such a high social temperature for a short period of time through wartime mobilization and other methods. However, if it is to be maintained for a long time, it may be achieved through some extraordinary means, such as seeking external economic assistance or even external expansion.

Of course, the rapid increase in internal pressure also means that it is less sensitive to changes in external pressure. Because even if the external pressure is as large as $p = 5\times10^4\,Pa$, there are still

$$p_i \gg p$$

This shows that the ability of the entire society to withstand pressure has increased. We can also calculate the change in entropy of the whole society:

$$\begin{aligned} S &= N\beta\left\{\frac{3}{2}lnT + \ln\left(\frac{V}{Nn_p}\right) + \frac{3}{2}\left[\frac{5}{3} + \ln\left(\frac{2\pi mk_B}{h^2}\right)\right]\right\} \\ &= 7.2\times10^9\times2.76\times10^4 \\ &\times\left\{\frac{3}{2}ln31800 + ln\frac{1.26\times10^{12}}{7.2\times10^9\times2\times10^{27}} + \frac{3}{2}\times\left[\frac{5}{3} + ln\left(\frac{2\pi mk_B}{h^2}\right)\right]\right\} \\ &\approx 5.01\times10^{15}\left(J\cdot K^{-1}\right) \end{aligned}$$

It can be seen that the entropy of the whole society has increased. This is because the change in entropy is directly related to the rise in social temperature. At the same time, the increase of the coupling coefficient of social interaction, if the concentration of the population can be greatly improved and the rise of social temperature can be suppressed, it is possible to reduce the entropy of the whole society. This is also possible, for example, if a social society successfully transforms into a family-based society, it can greatly reduce the social temperature. Of course, as a family society, it has its own problems that are difficult to solve.

3. The impact of rising global temperature on the integration of different cultures

In Section 9.3.2, we explored the "cultural junction" in a balanced society. That is a relatively unique phenomenon in a balanced society. When two different cultures come into contact, a "cultural junction" similar to the *PN* junction will be formed between the two cultures.

Of course, the formation of this kind of cultural junction is a non-equilibrium dynamic process, so it is difficult to deal with the equilibrium thermodynamic state equation. However, we can see from some important properties of semiconductor *PN* junctions that the same properties should also exist at the interface between two different cultures. That is, when the global temperature rises, it means that the ambient temperature of the cultural junction has risen. This rise in temperature will lead to a rise in the social temperature of the two different cultures. The rise in social temperature is conducive to the diffusion of social individuals.

Therefore, it can be expected that, in a very mild case, the process of integration between different cultures may be accelerated at the same time as the global temperature rises.

However, from the perspective of historical events, if some catastrophic events occur due to the rise in global temperature, there will be a transition and integration between cultures, and one culture will enter the area of another culture more deeply, which will lead to the emergence of strong repulsive force between different cultures, so there may be a variety of strong cultural conflicts. Violent exclusion of minorities is possible.

4. The impact of rising global temperatures on animal societies

Societies formed by animals are also an important factor in the entire Earth system. For artificially farmed animals, such as pigs, cattle, and sheep, their social temperature is very high. So these animals aren't much affected when global temperatures rise.

For example, for the current industrialized pigs, the density reaches 1 pig per square meter. So it can be calculated that for an area of 100 square meters, 100 pigs can be housed. Regardless of external pressure, the height of the pigs' social space is 2 meters, and the social interaction between pigs is $a_{pig} = 1.5 \times 10^9 Pa \cdot m^6$, so that the social temperature of the pig herd will reach:

$$T_{pig} = \frac{a_{pig} N}{\beta V} = \frac{1.5 \times 10^9 \times 100}{2.76 \times 10^4 \times 2 \times 100} \approx 28000K$$

It can be seen that this is a very high social temperature, mainly due to the high density of artificially farmed pigs. Therefore, the global temperature rises by a few degrees, or the temperature in a certain area rises by more than ten degrees. In fact, the impact on the artificially farmed pig herd is very small.

However, it will have a very large impact on animal societies living in the wild. Rising global temperatures have brought about faster plant growth. This gives wild animals at the bottom of the food chain more opportunities to reproduce. According to the Lotka-Volterra model (Wangersky, 1978), there are two populations of organisms, A and B, where A is the herbivores and B is the carnivores. Then there is such a relationship between the population numbers of the two

$$\begin{cases} \dfrac{dA}{dt} = \alpha A - \beta AB \\ \dfrac{dB}{dt} = \varepsilon \beta AB - \gamma B \end{cases} \tag{9-10}$$

It can be seen that if the global temperature increases, the coefficient α increases, which will lead to a rapid increase in the number of herbivores A. And the increase of A directly leads to the increase of B more through the second differential equation. So the whole wildlife world will become more prosperous.

It can also be seen from case study 5.2.4 that when the ambient temperature rises, it will help the wolves to obtain more energy. This aids in the reproduction of wolf pups, thereby increasing the number of wolves in the pack.

The rise in sea level will result in a substantial expansion of the ocean area. The ocean gets deeper. This certainly helps the growth of marine life. Therefore, marine animals will also usher in a more prosperous era.

References

Batchelor, C. L., Margold, M., Krapp, M., et al. (2019). The configuration of Northern Hemisphere ice sheets through the Quaternary. *Nature Communications*, 10, 3713.

EPA. (2022). *Overview of greenhouse gases*. United States Environmental Protection Agency. https://www.epa.gov/ghgemissions/overview-greenhouse-gases.

IPCC. (2022). Summary for policymakers, Special Report: Global Warming of 1.5 °C. https://www.ipcc.ch/sr15/chapter/spm/.

Jackson, P. (2014). *The Mongols and the West: 1221–1410*. Routledge.

Lacey, W. K. (1968). *The family in classical Greece*. Cornell University Press.

Morgan, D. (1986). *The mongols*. Oxford: Blackwell.

National Academy of Sciences. (2020). Climate change: Evidence and causes: Update 2020. Washington, DC: The National Academies Press. https://doi.org/10.17226/25733.

O'Brien, C. L., Robinson, S. A., Pancost, R. D., Damsté, J. S. S., Schouten, S., Lunt, D. J., ... & Wrobel, N. E. (2017). Cretaceous sea-surface temperature evolution: Constraints from TEX86 and planktonic foraminiferal oxygen isotopes. *Earth-Science Reviews*, 172, 224–247.

Patterson, C. B. (2009). *The family in Greek history*. Harvard University Press.

Rawson, B. (Ed.). (1987). *The family in ancient Rome: New perspectives*. Cornell University Press.

Tierney, J. E., Poulsen, C. J., Montañez, I. P., Bhattacharya, T., Feng, R., Ford, H. L., ... & Zhang, Y. G. (2020). Past climates inform our future. *Science*, 370(6517), eaay3701.

Trenberth, K. E. (1981). Seasonal variations in global sea level pressure and the total mass of the atmosphere. *Journal of Geophysical Research: Oceans*, 86(C6), 5238–5246.

Trenberth, K. E., & Guillemot, C. J. (1994). The total mass of the atmosphere. *Journal of Geophysical Research: Atmospheres*, 99(D11), 23079–23088.

Verniani, F. (1966). The total mass of the Earth's atmosphere. *Journal of Geophysical Research*, 71(2), 385–391.

Wangersky, P. J. (1978). Lotka-Volterra population models. *Annual Review of Ecology and Systematics*, 9, 189–218.

Appendix I:

Commonly used international system of units (SI) units

Symbol	Units	Meaning
Time	s	Second
Length	m	Meter
Mass	kg	Kilogram
Energy	J	Joule
Power	W	Watt
Volume	m^3	Cubic meter
Pressure	Pa	Pascal
Temperature	K	Kelvin
Amount of substance	mol	Mole
Force	N	Newton
Frequency	Hz	Hertz
Electricity	C	Coulomb

Appendix II:

Constants

1. Physical constants

Boltzmann constant: $k_B = 1.38 \times 10^{-23} J \cdot K^{-1}$
Avogadro constant: $N_A = 6.02 \times 10^{23}$
Planck's constant: $h = 6.63 \times 10^{-34} J \cdot s$
Water's molecular mass: $m_{water} = 3 \times 10^{-26} kg$
Van der Waals constant of water gas:
$a = 0.5537 Pa \cdot m^2 \cdot mol^{-2} = 1.53 \times 10^{-48} Pa \cdot m^6$
Critical temperature of water gas: $T = 647.30K$
Critical pressure of water gas: $P_c = 22.12MPa$
Critical volume of water gas: $v_c = 0.00317 m^3 kg^{-1} = 0.00000317 m^3 \cdot mol^{-1}$
The mass of the Earth's atmosphere: $m_{atm} = 5.14 \times 10^{18} kg$
Atmospheric heat capacity: $C_{atm} = 1006 Jkg^{-1}K^{-1}$
Earth's land area: $s = 1.5 \times 10^8 km^2 = 1.5 \times 10^{14} m^2$
Total area of Earth's surface including oceans:
$s_{total} = 5.1 \times 10^8 km^2 = 5.1 \times 10^{14} m^2$
The daily amount of energy radiated by the sun to the earth: $dE_{sun} = 1.5 \times 10^{22} J$

Steam turbine heat engine efficiency: approx. 25%
Internal combustion engine heat engine efficiency: approx. 40%

2. Some constants commonly used in this book

Human body volume: $v_p = 0.06m^3$
The number of water molecules contained in the individual: $n_p = 2\times 10^{27}$
Social constant: $\beta = 2.76\times 10^4 J\cdot K^{-1}$
Social interaction constant of social-oriented society: $a_s = 1.5\times 10^9 Pa\cdot m^6$
The social interaction constant of family-oriented society:
$a_{Hs} = 1.5\times 10^{10} Pa\cdot m^6$
The social interaction constant within the family: $a_H = 1.5\times 10^{11} Pa\cdot m^6$
Gender interaction constant: $a_g = 1.5\times 10^{12} Pa\cdot m^6$
During the farming era, the human body consumed energy per day: $dE_{nutrition} = 1200kcal$
The northern nomadic tribes consume energy per day: $dE_{nutrition} = 1400kcal$
The human body consumes energy per day in industrialized time: $dE_{nutrition} = 1500kcal$
Calories possessed by wheat: $dE_{wheat} = 3000kcal / kg$

Appendix III:

Parameters used in this book and their meaning

Parameters [1]	Meaning
p	Pressure
V	Volume
T	Temperature
μ	Chemical potential
E	Total energy
U	Internal energy
W	Work, or the number of states
Q	Heat
S	Entropy
H	Enthalpy, or the amount of information
F	Helmholtz free energy, or the force
G	Gibbs free energy, or the gravitational constant
γ	Cognitive Coupling Coefficient or Social Interaction Coupling Coefficient
g	Gender Interaction Coupling Coefficient
a	Van der Waals constant, which expresses the interaction between molecules
a_s	Social interaction constant of social-oriented society
a_{Hs}	The social interaction constant of family-oriented society

Parameters [1]	Meaning
a_H	The social interaction constant within the family
a_g	Gender interaction constant:
b	Van der Waals constant, which represents the social space of an individuals
v_p	Human body volume
β	Social constant. Converted from Boltzmann constant
n	Number of gas molecules
n_p	The number of water molecules possessed by the human body
N	Indicates the number of social individuals, or the amount of matter, in mol
h	Planck's constant
k_B	Boltzmann constant
N_A	Avogadro's constant
R	Gas constant
C	Heat capacity
C_v	Constant volume heat capacity
C_p	Constant pressure heat capacity
T_C	Critical social temperature
p_C	Critical social pressure
V_C	Critical social space
$\eta_{\alpha\beta}$	Minkowski space-time metric
$I(t)$	Number of neurons
$M(t)$	Cognitive pattern
$R(t)$	Network capacity
$C(t)$	Network complexity or Network depth
$D(t)$	Cognitive depth

[1] For parameters that do not appear in the above table, the descriptions in the relevant chapters of this book are used as the standard

Appendix IV:

Important formulas

Newton's law of gravitation

$$F = G\frac{Mm}{r^2}$$

Ideal gas equation of state

$$pV = nk_BT$$

Van der Waals equation

$$\left(p + \frac{n^2a}{V^2}\right)(V - b) = nk_BT$$

Physical system critical volume

$$V_C = 3b$$

Critical pressure

$$p_C = \frac{an^2}{27b^2}$$

Critical temperature

$$T_C = \frac{8an}{27k_B b}$$

Relationship between change in internal energy and change in heat and volume:

$$dU = dQ + dW = dQ - pdV$$

Relationship between heat and entropy (defined by Clausius):

$$dQ = TdS$$

The relationship between internal energy and temperature change

$$dU = CdT$$

The relationship between particle number, entropy, volume change and internal energy change

$$dU = TdS - pdV + \mu dN$$

Boltzmann statistical entropy calculation formula:

$$S = k_B lnW$$

Calculation formula for constant heat capacity

$$C_v = \frac{dQ}{dT} = \frac{dU}{dT}$$

Statistical formula for constant heat capacity
One-dimensional:

$$C_v = \frac{1}{2} N\beta$$

Two-dimensional:

$$C_v = N\beta$$

Three-dimensional:

$$C_v = \frac{3}{2} N\beta$$

The formula for calculating heat capacity at constant pressure:

$$C_p = \left(\frac{\partial H}{\partial T}\right)_p$$

Calculation formula of collision cross-section radius of space occupied by social individuals

$$r = \sqrt{\frac{4p}{\pi n f(T)}}$$

Ideal gas equation of state for social thermodynamic system:

$$pV = N\beta T$$

The Van der Waals equation of the social thermodynamic system:

$$\left(p + \frac{a_s N^2}{V^2}\right)(V - b) = N\beta T$$

Under the condition that there is no external pressure and the individual social space can be ignored, it can be simplified to:

$$a_s N = \beta T V$$

Charlie's law, there is a relationship under constant pressure:

$$\frac{V_1}{T_1} = \frac{V_2}{T_2}$$

Boyle's law, there is a relationship under constant temperature:

$$p_1 V_1 = p_2 V_2$$

Guy-Lussac's law, in the case of constant volume, there is a relationship:

$$\frac{P_1}{T_1} = \frac{P_2}{T_2}$$

The relationship between social individual internal energy and social temperature:

$$dU = C_v dT = N\beta dT$$

The condition of adiabatic process:

$$dQ = 0$$

Statistical formula for entropy of social thermodynamic system

$$S = \frac{3}{2} N\beta lnT + N\beta ln\frac{V}{Nn_p} + \frac{3}{2} N\beta\left[\frac{5}{3} + ln\left(\frac{2\pi mk_B}{h^2}\right)\right]$$

Helmholtz free energy

$$F = U - TS$$

Gibbs free energy

$$G = F + pV$$

Relationship between entropy and free energy:

$$S(T,V) = \left(\frac{\partial F}{\partial T}\right)_V$$

$$S(T,p) = \left(\frac{\partial G}{\partial T}\right)_p$$

The relationship between chemical potential and free energy:

$$\mu = \frac{G}{N}$$

Statistical formula for chemical potential of social thermodynamic system:

$$\mu(N,V,T) = \beta T ln\left[\frac{Nn_p}{V}\left(\frac{h^2}{2\pi mk_B T}\right)^{\frac{3}{2}}\right]$$

Relationships between enthalpy, free energy, and changes in temperature, volume, and social population.

$$dH = TdS + Vdp + \mu dN$$

$$dF = -SdT - pdV + \mu dN$$

$$dG = -SdT + Vdp + \mu dN$$

Criteria for judging entropy of equilibrium stability of social thermodynamic system

$$\delta S = 0$$

$$\delta^2 S < 0$$

Criteria for judging free energy

$$\delta F = 0$$

$$\delta^2 F > 0$$

$$\delta G = 0$$

$$\delta^2 G > 0$$

Critical parameters of a social thermodynamic system:

$$V_C = 3b$$

$$p_C = \frac{a_s N^2}{27b^2}$$

$$T_C = \frac{8a_s}{27\beta v_p}$$

www.ingramcontent.com/pod-product-compliance
Ingram Content Group UK Ltd.
Pitfield, Milton Keynes, MK11 3LW, UK
UKHW020838290926
15470UKWH00004B/1416